# AXIOMATIC PROJECTIVE GEOMETRY

# BIBLIOTHECA MATHEMATICA

*A series of Monographs on Pure and Applied Mathematics*
*Volume V*

*Edited with the cooperation of*

THE 'MATHEMATISCH CENTRUM'

and

THE 'WISKUNDIG GENOOTSCHAP'

at Amsterdam

*Editors:*

N. G. DE BRUIJN
J. DE GROOT
A. C. ZAANEN

# AXIOMATIC PROJECTIVE GEOMETRY

## Second edition

BY

## A. HEYTING

PROFESSOR EMERITUS OF MATHEMATICS
AT THE UNIVERSITY OF AMSTERDAM,
THE NETHERLANDS

1980

P. NOORDHOFF N.V. - GRONINGEN

NORTH-HOLLAND PUBLISHING COMPANY - AMSTERDAM

© NORTH-HOLLAND PUBLISHING COMPANY – 1980

WOLTERS-NOORDHOFF PUBLISHING

ISBN: 0 444 85431 2


*Publishers:*

NORTH-HOLLAND PUBLISHING COMPANY – AMSTERDAM · NEW YORK · OXFORD

WOLTERS-NOORDHOFF PUBLISHING – GRONINGEN


*Sole distributors for the U.S.A. and Canada:*

ELSEVIER NORTH-HOLLAND, INC.
52 VANDERBILT AVENUE
NEW YORK, N.Y. 10017


*First edition* 1963
*Second edition* 1980

**Library of Congress Cataloging in Publication Data**

Heyting, Arend, 1898–
Axiomatic projective geometry.

(Bibliotheca mathematica, a series of monographs
on pure and applied mathematics; v. 5)
Includes index.
1. Geometry, Projective—Foundations.
I. Title.   II. Series.
QA554.H48    1980    516.5    80-468
ISBN 0-444-85431-2


Printed in the Netherlands

# PREFACE

This book originates from a course of lectures which I have given several times in the University of Amsterdam. Since 1930 the subject has been revived by a number of mathematicians, of whom I mention Ruth Moufang, Marshall Hall Jr., Reinhold Baer and Günther Pickert. As one of the main results of their work the significance of weaker incidence propositions than that of Desargues was clarified. This part of the theory seems now sufficiently rounded off to be incorporated in an elementary textbook.

One of the main aims of the theory is the introduction of coordinates. Hereby a narrow connection between geometrical and algebraical investigations has been established, and algebraical theories have been suggested by geometrical problems. On the other hand, progress in geometry may be expected from the side of algebra. It is not the aim of this book to treat these algebraic questions extensively; its scope is more modest. I confined my attention to the direct consequences of various axioms for the geometry and for its set of coordinates. Therefore I preferred, where possible, geometrical to algebraical methods.

The book contains no new result of any importance, but I have tried to present the material in an easily accessible form. I hope that the book may help to renew the interest in geometry of a broad mathematical public.

It seemed unnecessary to give many references in a book like this: moreover, the excellent monograph by Günther Pickert, „Projektive Ebenen" (Springer 1955) contains a bibliography up to 1955. Of course most of the material can be found in Pickert's book; in particular the beautiful proof of theorem 2.4.6 has been borrowed from it.

The first chapter is introductory. Its first section contains a brief account of the axiomatic method; in the other sections I assembled some notions and theorems from diverse parts of mathematics, which I need in the rest of the book and which perhaps do not belong to the mathematical knowledge of the intended reader. It is advisable to skip after section 1.1 to

Chapter II and to return to the remaining sections of Chapter I when necessary.

I am grateful to the editors of ,,Bibliotheca Mathematica" for accepting the book in their series. My assistants W. Molenaar, A. Troelstra and F. Simons drew the figures and suggested many improvements; the latter also made the index and assisted in the reading of the proofs. Dr. and Mrs. de Vries kindly undertook a meticulous revision of the text from the linguistic point of view. I thank all these collaborators, and also the publishers, for their contributions to the final form of the book.

Amsterdam, January 1963.

A. HEYTING.

## PREFACE TO THE SECOND EDITION

Much work on projective planes has been done since 1963. For an account of this work I can refer to the books mentioned at the end of § 7,7 and to the literature mentioned in these books, for the elementary kernel of the subject remained the same. Therefore very little had to be changed in this edition; I only corrected some errors. In the appendix some additional information is given. I thank all those who drew my attention to corrections that could be made.

Some of the subjects treated in Chapter I belong to-day to the usual curriculum for undergraduates. I did not leave them out; any student can read as much of it as he needs.

I hope that the book will induce many students to a further study of geometry.

Castricum, October 1979.　　　　　　　　　A. HEYTING

# CONTENTS

# LIST OF SYMBOLS

# INTRODUCTION

## § 1.1. The axiomatic method.

1.1.1. The origin of mathematics was man's desire to systematize his knowledge. This tendency was particularly striking with the Greeks, and one of the culmination points of their scientific work was the systematization of geometry in Euclid's Elements. It is well known that under the hands of Greek mathematicians geometry had gradually taken the form of an axiomatic theory, which in Euclid reached its final shape. For many centuries his Elements were the model of a perfect mathematical theory; it was not before the nineteenth century that important corrections were made in them, which concerned the mathematical demonstrations as well as the interpretation of the axioms.

It is difficult to decide how the Greek philosophers and mathematicians conceived exactly the relation of their abstract geometry to ordinary space. One of the results of axiomatic work in the nineteenth century was the loosening of this connection. An axiom is no longer considered as an indubitable truth, and an axiomatic theory is but indirectly related with reality. Therefore not only geometry, but many other, even very abstract, mathematical theories have been axiomatized, and the axiomatic method has become a powerful tool for mathematical research, as well as a means of organizing the immense field of mathematical knowledge which thereby can be made more easily surveyable.

1.1.2. An axiomatic theory $\mathfrak{S}$ is described by giving
(i) a system of fundamental notions $\langle P_1, P_2, \ldots \rangle$,
(ii) a set of axioms about the fundamental notions.

We shall assume that the set of fundamental notions as well as the set of axioms is finite. The set of axioms of $\mathfrak{S}$ is also called its axiom system.

It will be useful to consider a simple example, which is not, like Euclidean geometry, charged with the misunderstandings caused by a millennial history.

The theory of groups ⑥ can be axiomatically described as follows. The set of fundamental notions is $\langle G, F \rangle$, where $G$ is a class and $F$ a function which assigns an element of $G$ to any ordered pair of elements of $G$ [in the language of set theory (§ 1.2): $F$ is a mapping of $G^2$ in $G$]. If $F$ assigns $c$ to the pair $(a, b)$, we write, as usual, $c = F(a, b)$.

The axioms of ⑥ are:

I. For any elements $a$, $b$, $c$ of $G$, $F(a, F(b, c)) = F(F(a, b), c)$.

II. For any elements $a$ and $b$ of $G$ there is a unique element $x$ of $G$ such that $F(a, x) = b$.

III. For any elements $a$ and $b$ of $G$ there is a unique element $y$ of $G$ such that $F(y, a) = b$.

I expresses the associative law, II and III the possibility and uniqueness of subtraction.

Group theory is the set of theorems which can be derived from the axioms; as we use in the proofs no other properties of the set $G$ and the function $F$ than those expressed in the axioms, $G$ and $F$ can be left unspecified. If we choose a particular set $G$ and a particular function $F$ (which, of course, must satisfy the axioms) we obtain a particular group. Every group, obtained by such a specification of $G$ and $F$, is a *model* of group theory. For instance, if we choose for $G$ the set $C$ of integers, with $F(a, b) = a+b$, the axioms are true. For this reason, $C$ together with the addition is a *model* of the theory of groups and of the axiom system (I, II, III). (The expressions "Model of an axiomatic theory $S$" and "model of the axiom system of $S$" will be used synonymously).

There exist many other models of the theory of groups. In § 3.4 we shall consider models for projective geometry. I shall not give a general definition of a model for an axiomatic theory [see e.g. A. Tarski, Introduction to logic and to the Methodology of deductive Sciences, Ch. VI (New York, 1939)]; the example above will suffice to explain the notion.

Some models of group theory are mentioned in the exercises. The plurality of models has important consequences for the purport of theorems in the axiomatic theory. Let us consider an example. It is easy to derive from I, II, III the theorem

($T$) There is an element $e$ of $G$ such that, for every element $a$ of $G$, $F(a, e) = F(e, a) = a$, is true.

In the model of the integers, ($T$) expresses the fundamental property of the number 0. In other models it will represent different properties. By proving ($T$) we prove all these properties at the same time.

In general, if we derive a theorem $T$ in an axiomatic theory $\mathfrak{S}$, we have also proved the theorems which correspond to $T$ in all the models of $\mathfrak{S}$. In this way the axiomatic method allows us to economize enormously on mathematical proofs. Moreover, in recognizing different theories as models of one axiomatic theory, we obtain new information about the relations between the former.

But there is still another use of the method, which shows new advantages. It may happen that for the proof of a theorem we do not need all the axioms, but only some of them. Such a theorem is true not only for models of the whole system, but also for those of the smaller system which contains only the axioms used in the proof. Thus it is important in an axiomatic theory to prove every theorem from the least possible set of axioms. This point of view is prominent in the axiomatics of geometry.

Historically, at least some of the models of an axiomatic theory precede this theory.

It would be interesting, but outside the scope of this book, to follow the history of the theory of groups from its initial state, in which every model was studied separately, up to its modern axiomatic form. The history of Euclidean geometry is somewhat different because the axiomatic method was applied to it so very early, when this method had not yet been fully developed, so that its advantages could not be completely realized. Hilbert [Grundlagen der Geometrie; first edition 1899] was the first to detach it completely from its original model, the theory of ordinary space. Thereby he made it into a part of pure mathematics in the modern sense. The axiomatic theory must be fit for application to any of its models, even to one that has nothing to do with space. Therefore space-intuition can play no part in the deductions. This condition was not satisfied by Euclid's system; for instance, Euclid uses freely in his proofs several properties of the betweenness relation, which he does not mention in his axioms, so that no other foundation than space-intuition is available for them. The task of liberating geometry from these ties with space-intuition was fulfilled between 1860 and 1900 by a number of mathematicians, among whom Pasch and Hilbert were prominent.

Perhaps it seems astonishing that even after this liberation figures occur in books on geometry. Indeed they are inessential, and if the book is good they can be left out without loss of coherence. Their main use is to support the memory and to facilitate the comprehension of complicated proofs. On the other hand, they always threaten to seduce us into illicit use of space-intuition; we must be on our guard against this danger which has caused mistakes even in excellent works on axiomatics.

If we are asked what an axiom is in modern mathematics, perhaps the best answer is: a starting-point for deductions. As explained above, the theorems obtained by deduction from an axiom system $A$ are valid for every model of $A$.

**1.1.3.** It is necessary to distinguish between *propositions* and *theorems*. A *proposition* in an axiomatic theory $\mathfrak{S}$ is a sentence in which no other notions occur than the fundamental notions of $\mathfrak{S}$ and logical notions. This definition suffices for our purposes; a more precise definition would be possible after a detailed description of the formal structure of an axiomatic theory. It is essential to remark that a proposition in $\mathfrak{S}$ need not be valid in $\mathfrak{S}$. A proposition is *valid* in $\mathfrak{S}$ if it is an axiom or if it can be deduced from the axioms. A valid proposition is also called a *theorem* of $\mathfrak{S}$. Every theorem of $\mathfrak{S}$ is true for every model of $\mathfrak{S}$, but a proposition in $\mathfrak{S}$ can be true for a certain model of $\mathfrak{S}$, without being a theorem of $\mathfrak{S}$. For instance, if $\mathfrak{S}$ is group theory, the proposition which expresses the commutative property is not a theorem of $\mathfrak{S}$, but it is true in the model of the integers under addition. We shall not discuss the question whether a proposition in $\mathfrak{S}$, which is true for every model of $\mathfrak{S}$, must be a theorem of $\mathfrak{S}$.

**1.1.4.** Let us summarize. The characteristic features of an axiomatic theory are the following.

1. A complete list of the fundamental notions of the theory is given.
2. Every other notion is reduced to the fundamental notions by explicit definition.
   These definitions must be of such a nature, that everywhere, except in the definition itself, the definiens can be substituted for the definiendum. Consequently, we could in principle dispense with the defined notions.
3. A complete list of fundamental theorems (called axioms) is given.
4. Every other theorem is deduced from the axioms by logical reasoning.

1.1.5.  In elementary axiomatics, logic is used in an unanalysed form. The analysis of logic is one of the main subjects of the investigations on the foundations of mathematics. However, in this book we shall take the elementary point of view and accept the validity of the usual (two-valued) logic.

Without going into a general theory of axiomatic theories, we must mention some conditions to which an axiomatic theory can be subjected.

1.1.6.  **Consistency.** An axiomatic theory is *consistent* if no contradiction can be derived in it. It is clear that an inconsistent theory can have no model; therefore it is of utmost importance to know that our axiomatic theories are consistent.

The simplest way to prove the consistency of an axiomatic theory is to provide a model for it. Thus group theory is consistent because the integers under addition form a model for it. Strictly speaking, a consistency proof by the model method is relative; we have only proved that group theory is consistent if the arithmetic of integers is consistent. The question whether and how absolute consistency proofs are possible, is another main topic in investigations on the foundations of mathematics. Here, taking the elementary point of view, we shall admit the arithmetic of real numbers as sufficiently safe and use it freely for the construction of models.

1.1.7.  **Independence.** An axiom $A$ is **independent of** a set of axioms $B$ if it cannot be derived from $B$. An axiom system is *independent* if every one of its axioms is independent of the remaining ones. Of course, in case an axiom system is not independent, so that one of its axioms, say $A$, can be derived from the others, then $A$ is redundant in this sense that the same theorems can be derived from the complete system as from that which remains after leaving $A$ out.

The independence of an axiom system can be proved by the model method, as follows. In order to prove that $A$ is independent of the set of axioms $B$ it suffices to construct a model of $B$ in which $A$ is not valid. Indeed, if $A$ could be derived from $B$, then $A$ would be valid in every model of $B$.

**1.1.8. Completeness.** In many cases an axiomatic theory $\mathfrak{S}$ is constructed in connection with a theory $\mathfrak{T}$ which existed before the axiomatization. Such is the case for projective geometry. One may then ask whether $\mathfrak{S}$ is sufficient to derive all the theorems of $\mathfrak{T}$; if this is the case, $\mathfrak{S}$ is said to be *complete* with respect to $\mathfrak{T}$. In order to make this notion precise it is necessary to describe $\mathfrak{T}$ more accurately than it is generally done. We shall come back to this question later (§ 2.2).

**1.1.9. Categoricity.** An axiomatic system is *categorical* if any two of its models are isomorphic. I prefer to explain the notion of isomorphism in the concrete case of projective geometry (see § 2.1).

**1.1.10. Definitions.** The axiomatic theory $\mathfrak{S}_1$ is *contained* in the axiomatic theory $\mathfrak{S}_2$ ($\mathfrak{S}_1 \subseteqq \mathfrak{S}_2$), if the following two conditions are satisfied:

I. Every fundamental notion of $\mathfrak{S}_1$ is also a fundamental notion of $\mathfrak{S}_2$.

II. Every theorem of $\mathfrak{S}_1$ is also a theorem of $\mathfrak{S}_2$.

$\mathfrak{S}_1$ is *properly* contained in $\mathfrak{S}_2$ and $\mathfrak{S}_2$ is a *proper extension* of $\mathfrak{S}_1$($\mathfrak{S}_1 \subset \mathfrak{S}_2$), if $\mathfrak{S}_1$ is contained in $\mathfrak{S}_2$ and at least one theorem of $\mathfrak{S}_2$ is not a theorem of $\mathfrak{S}_1$.

REMARK. We obtain a wider notion of a theory contained in another theory, if we read instead of I:

Every fundamental notion of $\mathfrak{S}_1$ can be defined in $\mathfrak{S}_2$. This notion will not be used in the sequel.

Evidently, a sufficient condition for $\mathfrak{S}_1 \subseteq \mathfrak{S}_2$ is that I holds and every axiom of $\mathfrak{S}_1$ is also an axiom of $\mathfrak{S}_2$. This condition is not necessary, because two different axiom systems can lead to the same set of theorems.

### 1.1.11. Exercises.

1. Show that the following theories are models of the axiomatic system for group theory, as described in 1.1.2.
   a. $G$ is the set of rationals $\neq 0$; $F(a, b)$ is $ab$.
   b. $G$ is the set of congruent transformations in the plane; $F(a, b)$ is the transformation which results by applying first the transformation $a$, then the transformation $b$.
2. Why is the example in exercise 1 a) not a model for group theory if we omit the condition $\neq 0$?
3. Prove $(T)$ (1.1.2) from I, II, III, or find the proof in a book on group theory.
4. Give as many models as you can find for the following axiomatic theory:
   Fundamental notions: $\langle C, O \rangle$, where $C$ is a class and $O$ a binary relation in $C$.
   Axioms: (i). For any two elements $a$, $b$ of $C$, $O(a, b)$ or $O(b, a)$ or both.
   (ii) If $O(a, b)$ and $O(b, a)$, then $a = b$.
   (iii) If $O(a, b)$ and $O(b, c)$, then $O(a, c)$.
   Is there a model in which $C$ is a class of one element?
5. Is the axiomatic theory, described in exercise 4, consistent?
6. Prove that it becomes inconsistent if we add the axioms (iv) and (v):
   (iv) $C$ contains two different elements.
   (v) For any two different elements $a$ and $b$ of $C$, $O(a, b)$ is true.
7. A consistent theory remains consistent if we omit one of the axioms.
8. Prove that in the axiom system for group theory (1.1.2), I is independent of II and III.
   [Hint. Think of subtraction instead of addition].

9. Is group theory complete with respect to the theory of addition of integers?

   [Answer: No, for the commutative law is independent of I, II, III. See exercise 1b].

10. The axiomatic theory $\mathfrak{S}_1$ is contained in the axiomatic theory $\mathfrak{S}_2$ if (I) every fundamental notion of $\mathfrak{S}_1$ is also a fundamental notion of $\mathfrak{S}_2$ and (II) every axiom of $\mathfrak{S}_1$ is a theorem of $\mathfrak{S}_2$.

11. Let $\mathfrak{S}_1$ be the theory of groups, as defined above (1.1.2) and $\mathfrak{S}_2$ the theory which results from $\mathfrak{S}_1$ by adding the axiom: "For any elements $a$, $b$ of $G$, $F(a, b) = F(b, a)$." Show that $\mathfrak{S}_1 \subset \mathfrak{S}_2$.

## Preliminaries.

In the following sections I assemble some notions and theorems from other parts of mathematics which are used in the body of the book. Most of the proofs are omitted. Of course, it is not necessary to read this before the rest of the book; the reader can reach back to it wherever he needs it.

## § 1.2. Notions from set theory.

The fundamental notions of set theory are used throughout the book; order relations occur in Chapter VII.

The reader is assumed to be familiar with the notions of sum (union) and meet (intersection) of sets, and of set-inclusion. The union of sets $A$ and $B$ is denoted by $A \cup B$, their intersection by $A \cap B$. If $A$ is a subset of $B$ we write $A \subseteq B$; $A \subset B$ means that $A$ is a proper subset of $B$, i.e. $A \subseteq B$ and $B$ contains at least one element not in $A$. $a \in A$ means: $a$ is an element of the set $A$.

The *direct product* $A \times B$ of sets $A$ and $B$ is the set of ordered pairs $(a, b)$, where $a \in A$, $b \in B$.

A binary *relation* $R$ between $A$ and $B$ is a subset of $A \times B$. If $(a, b) \in R$, we also say that $a$ and $b$ are in the relation $R$, or that $aRb$. A relation in $A$ is a relation between $A$ and $A$.

A *mapping* of $A$ into $B$ is a relation $R$ between $A$ and $B$ such that for every element $a$ of $A$ there is exactly one element $b$ of $B$ satisfying $aRb$.

The mapping $R$ is *onto* $B$, if for every element $b$ of $B$ there is at least one element $a$ of $A$ satisfying $aRb$.

The mapping $R$ of $A$ onto $B$ is *one-to-one* if for every element $b$ of $B$ there is exactly one element $a$ of $A$ satisfying $aRb$.

A relation $R$ in $A$ is *symmetric*, if $aRb$ implies $bRa$. It is *transitive* if $aRb$ and $bRc$ imply $aRc$. It is *reflexive* if $aRa$ for every $a$ in $A$. A reflexive, symmetric and transitive relation is called an *equivalence relation*.

With respect to an equivalence relation $R$ in $A$, $A$ can be divided into *equivalence classes* such that (I) every element of $A$ belongs to exactly one equivalence class, and (II) two elements $a$ and $b$ are in the relation $R$ ($aRb$) if and only if they are in the same equivalence class.

The relation $R$ in $A$ is an *order relation* if it is transitive, and satisfies the conditions (I) $aRb$ implies $a \neq b$, and (II) for any two different elements $a$ and $b$ of $A$, either $aRb$ or $bRa$ holds. We also say that $A$ is *ordered* by the relation $R$. An order relation is often denoted by $<$.

In the set $A$, ordered by $<$, $c$ is *between* $a$ and $b$, if either $a < c < b$ or $b < c < a$.

Of any three elements of an ordered set exactly one is between the other two. The set of elements of $A$ which are between $a$ and $b$, is the *segment* $(a, b)$. Four different elements $a$, $b$, $c$, $d$ of an ordered set can be divided in exactly one way into *alternating pairs*. For instance, $a$, $b$ and $c$, $d$ are alternating pairs if one of $c$, $d$ is between $a$, $b$ and the other not; then also one of $a$ and $b$ is between $c$ and $d$, and the other not. Thus the relation between alternating pairs is symmetric. The relation between alternating pairs defines a *cyclical order* in $A$. This notion will be discussed in detail in § 7.1.

If $A$ and $B$ are disjoint ordered sets, then the *ordered* sum $A+B$ of $A$ and $B$ is defined as follows. Let $<\limits_A$, $<\limits_B$ be the order relations in $A$, $B$ respectively. $A+B$ is the union $A \cup B$, in which an order relation $<$ is given by the rules (I), (II), (III):

(I) If $x$, $y \in A$, then $x < y$ if and only if $x <\limits_A y$.

(II) If $x$, $y \in B$, then $x < y$ if and only if $x <\limits_B y$

(III) If $x \in A$, $y \in B$, then $x < y$.

It is easily verified that $<$ is an order relation in $A \cup B$.

Analogously the ordered sum of an ordered finite, or even infinite, set of ordered sets can be defined.

An $n$-ary relation in $A$ is a subset of $A^n$.

## § 1.3. Notions from algebra.

Ordered groups and fields occur in Chapter VII.

Quaternions and Cayley's algebra are used for the construction of models in § 3.5.

1.3.1. The notions of a group, an abelian (commutative) group, a division ring (skew field), a field, are assumed to be known, as well as that of isomorphism between groups, and between fields.[1])

A group $G$ in which the group operation is denoted by $\circ$, is an *ordered group* if it is ordered and, for any elements $a$, $b$, $c$ of $G$, $a < b$ entails $a \circ c < b \circ c$ and $c \circ a < c \circ b$.

A division ring $K$ is an *ordered* division ring if it is ordered, its additive group is an ordered group, and moreover $a > 0$ and $b > 0$ entail $ab > 0$. The field of rationals and the field of reals are ordered fields, if ordered in the usual way.

1.3.2. **Quaternions.** The simplest example of a division ring which is not a field is the division ring $Q$ of *real quaternions*.

$Q$ is the set of quadruples $(a_0, a_1, a_2, a_3)$ of real numbers with the composition rules:

$(a_0, a_1, a_2, a_3) + (b_0, b_1, b_2, b_3) = (a_0 + b_0, a_1 + b_1, a_2 + b_2, a_3 + b_3)$.
$(a_0, a_1, a_2, a_3) \ (b_0, b_1, b_2, b_3) =$
$(a_0 b_0 - a_1 b_1 - a_2 b_2 - a_3 b_3, \ a_0 b_1 + a_1 b_0 + a_2 b_3 - a_3 b_2,$
$a_0 b_2 + a_2 b_0 + a_3 b_1 - a_1 b_3, \ a_0 b_3 + a_3 b_0 + a_1 b_2 - a_2 b_1)$

The elements $(1, 0, 0, 0)$, $(0, 1, 0, 0)$, $(0, 0, 1, 0)$ and $(0, 0, 0, 1)$ are denoted by $e$, $j_1$, $j_2$, $j_3$ respectively; then $(a_0, a_1, a_2, a_3) = a_0 e + a_1 j_1 + a_2 j_2 + a_3 j_3$.

For the elements $e$, $j_1$, $j_2$, and $j_3$ the following multiplication table holds:

---

[1]) We shall use the word "field" in the sense of "commutative division ring".

|   | $e$ | $j_1$ | $j_2$ | $j_3$ |
|---|---|---|---|---|
| $e$ | $e$ | $j_1$ | $j_2$ | $j_3$ |
| $j_1$ | $j_1$ | $-e$ | $j_3$ | $-j_2$ |
| $j_2$ | $j_2$ | $-j_3$ | $-e$ | $j_1$ |
| $j_3$ | $j_3$ | $j_2$ | $-j_1$ | $-e$ |

,

which shows that $Q$ is non-commutative.

The product $(a_0e+a_1j_1+a_2j_2+a_3j_3)(b_0e+b_1j_1+b_2j_2+b_3j_3)$ can now be computed by using the distributive laws and applying the multiplication table, where the units $e$, $j_1$, $j_2$, $j_3$ commute with real numbers.

It can be verified by direct computation that the axioms for a division ring hold in $Q$. For those who are acquainted with the theory of matrices, an easy access to the theory of quaternions is by two-rowed complex matrices. Denote by $E$, $J_1$, $J_2$, $J_3$ respectively the matrices

$$\begin{pmatrix}1 & 0 \\ 0 & 1\end{pmatrix}, \begin{pmatrix}i & 0 \\ 0 & -i\end{pmatrix}, \begin{pmatrix}0 & 1 \\ -1 & 0\end{pmatrix}, \begin{pmatrix}0 & i \\ i & 0\end{pmatrix};$$

then the same multiplication table as for $e$, $j_1$, $j_2$, $j_3$ applies to $E$, $J_1$, $J_2$, $J_3$.

The matrices of the form

$$A = a_0E+a_1J_1+a_2J_2+a_3J_3 = \begin{pmatrix} a_0+a_1i & a_2+a_3i \\ -a_2+a_3i & a_0-a_1i \end{pmatrix}$$

form a subring $S$ of the ring of two-rowed complex matrices. As

$$\det A = \begin{vmatrix} a_0+a_1i & a_2+a_3i \\ -a_2+a_3i & a_0-a_1i \end{vmatrix} = a_0^2+a_1^2+a_2^2+a_3^2,$$

every element of $S$ except the null-matrix has an inverse; consequently $S$ is a division ring.

It follows easily from the correspondence between the multiplication tables, that the mapping $\varphi$ with $\varphi(a_0e+a_1j_1+a_2j_2+a_3j_3)=$ $=a_0E+a_1J_1+a_2J_2+a_3J_3$ is an isomorphism between $S$ and $Q$. Thus $Q$ is a division ring.

The *conjugate* of the quaternion $a = a_0e+a_1j_1+a_2j_2+a_3j_3$ is $\bar{a} = a_0-a_1j_1-a_2j_2-a_3j_3$. It is easily verified, either directly or by means of the matrix representation, that

$$a\bar{a} = \bar{a}a = (a_0^2+a_1^2+a_2^2+a_3^2, 0, 0, 0).$$

The real number $a_0{}^2+a_1{}^2+a_2{}^2+a_3{}^2$ is called the *norm* of $a$ and denoted by $N(a)$.

For the conjugate of a product we have the rule $\overline{ab} = \bar{b}\bar{a}$. It follows that $N(ab) = (ab)\overline{ab} = ab\bar{b}\bar{a} = aN(b)\bar{a} = a\bar{a}N(b) = N(a)N(b)$.

The quaternions $(a_0, 0, 0, 0)$ form a subfield of $Q$, which is isomorphically mapped on the field of reals by the mapping $(a_0, 0, 0, 0) \to a_0$. If we identify corresponding elements in this mapping, the field of reals becomes a subfield of $Q$. We may now write $a\bar{a} = N(a)$.

### 1.3.3.  Cayley's algebra.

Cayley's algebra $C$ is a set in which an addition and a multiplication are defined. It is an abelian group with respect to addition, the distributive laws hold, and left- and right-hand division, except by 0, are always possible and unique, but the associative law is not generally valid. The following method is due to Dickson [Trans. Amer. Math. Soc. **13** (1912)].

$C$ can be described as the set of pairs $(\alpha_1, \alpha_2)$, where $\alpha_1$ and $\alpha_2$ are real quaternions, with addition defined by

$$(1) \qquad (\alpha_1, \alpha_2)+(\beta_1, \beta_2)=(\alpha_1+\beta_1, \alpha_2+\beta_2)$$

and multiplication by

$$(2) \qquad (\alpha_1, \alpha_2)(\beta_1, \beta_2)=(\alpha_1\beta_1-\bar{\beta}_2\alpha_2, \beta_2\alpha_1+\alpha_2\bar{\beta}_1).$$

The distributive laws are easily verified; they are immediate consequences of the fact that the right member of (2) is linear in each of $(\alpha_1, \alpha_2)$ and $(\beta_1, \beta_2)$. In order to prove the possibility and uniqueness of right-hand division, we have to solve the equations for $\xi_1$, $\xi_2$:

$$\alpha_1\xi_1-\bar{\xi}_2\alpha_2 = \beta_1, \qquad \text{(i)}$$
$$\xi_2\alpha_1+\alpha_2\bar{\xi}_1 = \beta_2. \qquad \text{(ii)}$$

Multiply (i) on the left by $\bar{\alpha}_1$:

$$\bar{\alpha}_1\alpha_1\xi_1-\bar{\alpha}_1\bar{\xi}_2\alpha_2 = \bar{\alpha}_1\beta_1. \qquad \text{(iii)}$$

From (ii) we have $\xi_2\alpha_1 = \beta_2-\alpha_2\bar{\xi}_1$, so, taking the conjugate, $\bar{\alpha}_1\bar{\xi}_2 = \bar{\beta}_2-\xi_1\bar{\alpha}_2$.

Substitute in (iii):

$$\bar{\alpha}_1\alpha_1\xi_1-\bar{\beta}_2\alpha_2+\xi_1\bar{\alpha}_2\alpha_2 = \bar{\alpha}_1\beta_1.$$

Noting that $\bar{\alpha}_2\alpha_2$ is a real number, so that $\xi_1\bar{\alpha}_2\alpha_2 = \bar{\alpha}_2\alpha_2\xi_1$, we have finally $(\bar{\alpha}_1\alpha_1+\bar{\alpha}_2\alpha_2)\xi_1 = \bar{\alpha}_1\beta_1+\bar{\beta}_2\alpha_2$. This equation allows us to compute $\xi_1$.

In order to find $\xi_2$, multiply (ii) by $\bar{\alpha}_1$ on the right:

$$\xi_2\alpha_1\bar{\alpha}_1+\alpha_2\bar{\xi}_1\bar{\alpha}_1 = \beta_2\bar{\alpha}_1 \qquad \text{(iv)}$$

From (i): $\alpha_1\xi_1 = \beta_1+\bar{\xi}_2\alpha_2$; $\bar{\xi}_1\bar{\alpha}_1 = \bar{\beta}_1 + \bar{\alpha}_2\xi_2$.
Substitute in (iv):

$$\xi_2\alpha_1\bar{\alpha}_1 + \alpha_2(\bar{\beta}_1+\bar{\alpha}_2\xi_2) = \beta_2\bar{\alpha}_1$$
$$(\alpha_1\bar{\alpha}_1+\alpha_2\bar{\alpha}_2)\xi_2 = \beta_2\bar{\alpha}_1-\alpha_2\bar{\beta}_1.$$

From this equation we compute $\xi_2$.

In particular, let us take $\beta = 1$, so that $\beta_1 = 1, \beta_2 = 0$. Denoting $\alpha_1\bar{\alpha}_1+\alpha_2\bar{\alpha}_2$ by $N\alpha$, we find (provided $\alpha \neq 0$):

$$(\xi_1, \xi_2) = \frac{1}{N\alpha}(\bar{\alpha}_1, -\alpha_2).$$

Thus $(\alpha_1, \alpha_2)(\bar{\alpha}_1, -\alpha_2) = N\alpha$.

As the relation between $(\alpha_1, \alpha_2)$ and $(\bar{\alpha}_1, -\alpha_2)$ is symmetric, we have also $(\bar{\alpha}_1, -\alpha_2)(\alpha_1, \alpha_2) = N\alpha$.

This leads us to the denotation $(1/N\alpha)(\bar{\alpha}_1, -\alpha_2) = \alpha^{-1}$, so that $\alpha\alpha^{-1} = \alpha^{-1}\alpha = 1$.

The reader may now verify the following special cases of the associative law for multiplication:

$$(\beta\alpha)\alpha^{-1} = \beta. \qquad \alpha^{-1}(\alpha\beta) = \beta.$$

The following example shows that the associative law for multiplication is not generally valid.

$$\{(j_1, 0)(j_2, 0)\}(0, j_1) = (0, -j_2),$$
$$(j_1, 0)\{(j_2, 0)(0, j_1)\}=(0, j_2);$$

The verification is left to the reader.

## § 1.4. Analytic projective geometry.

*APPG* is used throughout the book as the standard model for the axiom system of plane projective geometry; *ASPG* from chapter **IV** onwards for that of threedimensional projective geometry.

The reader may have some knowledge of projective geometry,

obtained on the basis of Euclidean geometry, or by some other approach. I shall sketch here a method which uses only elementary algebra and which presupposes no geometrical knowledge. It is well known that in plane projective geometry coordinates can be introduced, such that a point is determined by the ratio $(x_0, x_1, x_2)$ of three numbers. In fact, we use this as the definition of a point. In this section we use "number" in the sense of "real number"; however, the theory remains valid for any field. Later on it will be extended to arbitrary division rings (§ 3,4). Let $F$ denote the field of reals; $F^n$ is the set of sequences of $n$ numbers $(x_1, \ldots, x_n)$.

**Definition.** Two sequences of $n$ numbers $(x_1, \ldots, x_n)$ and $(y_1, \ldots, y_n)$ are *proportional* if there exists a number $c \neq 0$, such that $y_i = cx_i$ $(i = 1, \ldots, n)$.

It is easy to see that proportionality is an equivalence relation in $F^n$. Therefore $F^n$ can be divided into equivalence classes with respect to this relation. The null sequence $(0, \ldots, 0)$ constitutes such an equivalence class.

**Definition.** An equivalence class in $F^n$ with respect to the relation of proportionality, which does not contain the null sequence, is a *ratio* (more exactly: an $n$-ratio).

In other words: Two sequences $(x_1, \ldots, x_n)$ and $(y_1, \ldots, y_n)$ are in the same ratio, if they are proportional.

We can now define analytic plane projective geometry ($APPG$).

**Definition.** A point of $APPG$ is a ratio of 3 numbers $(x_0, x_1, x_2)$ . $x_0, x_1, x_2$ form a set of coordinates of the point. It is clear that $(cx_0, cx_1, cx_2)$ $(c \neq 0)$ is another set of coordinates of the same point.

**Definition.** A line of $APPG$ is the set of points whose coordinates satisfy a linear equation:

$$(1) \qquad a_0x_0 + a_1x_1 + a_2x_2 = 0 \quad (a_0, a_1, a_2 \text{ not all } 0).$$

Note that the line is determined by the ratio $(a_0, a_1, a_2)$, and conversely. Therefore $(a_0, a_1, a_2)$ is called a set of coordinates of the line. Again $(ca_0, ca_1, ca_2)$ $(c \neq 0)$ is another set of coordinates of the same line.

It is well known from linear algebra that, if $(p_0, p_1, p_2)$ and $(q_0, q_1, q_2)$ are non-proportional solutions of (1), then all the solutions of (1) are obtained by substituting real numbers for $\lambda$ and $\mu$ in $x_i = \lambda p_i + \mu q_i$ $(i = 0, 1, 2)$. This gives

**Theorem 1.4.1.** If $P(p_0, p_1, p_2)$ and $Q(q_0, q_1, q_2)$ are different points, then a parametric representation of the line $PQ$ is $x_i = \lambda p_i + \mu q_i$ $(i = 0, 1, 2)$; abbreviated: $X = \lambda P + \mu Q$.

**Theorem 1.4.2.** If $a_0 x_0 + a_1 x_1 + a_2 x_2 = 0$ and $b_0 x_0 + b_1 x_1 + b_2 x_2 = 0$ are lines, intersecting in a point $S$, then the equation of any line through $S$ is of the form

$$(2) \qquad \lambda(a_0 x_0 + a_1 x_1 + a_2 x_2) + \mu(b_0 x_0 + b_1 x_1 + b_2 x_2) = 0.$$

PROOF. It is clear that (2) represents a line through $S$. Now let $l$ be a line through $S$ and $P(p_0, p_1, p_2)$ a point on $l$, $P \neq S$. We can find $\lambda_1$ and $\mu_1$ such that

$$\lambda_1(a_0 p_0 + a_1 p_1 + a_2 p_2) + \mu_1(b_0 p_0 + b_1 p_1 + b_2 p_2) = 0.$$

Then $\lambda_1(a_0 x_0 + a_1 x_1 + a_2 x_2) + \mu_1(b_0 x_0 + b_1 x_1 + b_2 x_2) = 0$ is the equation of the line $SP$, that is $l$.

If $\alpha$ and $\beta$ are lines, we shall often denote their equations by $\alpha = 0$ and $\beta = 0$; the line with equation $\lambda\alpha + \mu\beta = 0$ will be denoted by $\lambda\alpha + \mu\beta$.

In the preceding definitions and theorems a peculiarity of $APPG$ becomes apparent, namely the duality principle, which consists in the possibility of interchanging points and lines in the theorems. I shall not go into this subject here, because I have to treat it in the main text of the book (see Th. 3.3.2). I need only a few theorems of $APPG$.

**Theorem 1.4.3.** Given two different points, there is exactly one line which passes through both of them.

PROOF. Let $P(p_0, p_1, p_2)$ and $Q(q_0, q_1, q_2)$ be different points. The equations $\xi_0 p_0 + \xi_1 p_1 + \xi_2 p_2 = 0$ and $\xi_0 q_0 + \xi_1 q_1 + \xi_2 q_2 = 0$ have exactly one ratio $(\xi_0, \xi_1, \xi_2)$ as a solution. This is the set of coordinates of the required line.

REMARK. The equation of $PQ$ is

$$\begin{vmatrix} x_0 & x_1 & x_2 \\ p_0 & p_1 & p_2 \\ q_0 & q_1 & q_2 \end{vmatrix} = 0.$$

It is easily seen by substitution that the coordinates of $P$, and those of $Q$ satisfy this equation.

**Theorem 1.4.4.** Given two different lines, there is exactly one point which belongs to both of them.

**PROOF.** The equations of the lines, $a_0x_0+a_1x_1+a_2x_2 = 0$ and $b_0x_0+b_1x_1+b_2x_2 = 0$, determine a ratio $(x_0, x_1, x_2)$ as their solution. This is the set of coordinates of the point of intersection.

**Theorem 1.4.5.** (Desargues' theorem). Let $A_1A_2A_3$ and $B_1B_2B_3$ be triangles; $A_i$ and $B_i$ $(i = 1, 2, 3)$ are called corresponding vertices; sides connecting corresponding vertices, e.g. $A_1A_2$ and $B_1B_2$, are corresponding sides. It is supposed that corresponding

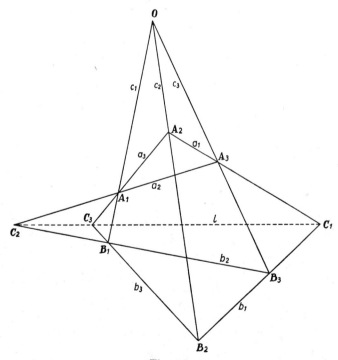

Fig. 1.1.

vertices as well as corresponding sides are different. If the lines connecting corresponding vertices pass through a point $O$, then the points of intersection of corresponding sides are on a line.

Several proofs are given in books on projective geometry. I give here a purely algebraic proof.

**Proof.** By Th. 1.4.1 we have $B_1 = \rho O + \sigma A_1$; we may divide the coordinates of $B_1$ by $\rho$, so that

$$B_1 = O + \lambda_1 A_1.$$

Similarly,
$$B_2 = O + \lambda_2 A_2,$$
$$B_3 = O + \lambda_3 A_3.$$

If $A_2 A_3$ intersects $B_2 B_3$ in $C_1$, then

$$C_1 = \mu_2 A_2 + \mu_3 A_3,$$

but also

$$C_1 = \nu_2 B_2 + \nu_3 B_3 = \nu_2 (O + \lambda_2 A_2) + \nu_3 (O + \lambda_3 A_3)$$
$$= (\nu_2 + \nu_3) O + \nu_2 \lambda_2 A_2 + \nu_3 \lambda_3 A_3.$$

This gives $(\nu_2 + \nu_3) O = (\mu_2 - \nu_2 \lambda_2) A_2 + (\mu_3 - \nu_3 \lambda_3) A_3$; so if $\nu_2 + \nu_3 \neq 0$, $O$ would be on $A_2 A_3$, which is not the case. It follows that $\nu_2 + \nu_3 = 0$.

Thus $C_1 = \nu_2 (\lambda_2 A_2 - \lambda_3 A_3)$.

After dividing the coordinates of $C_1$ by $\nu_2$, we obtain

$$C_1 = \lambda_2 A_2 - \lambda_3 A_3.$$

Similarly, if $C_2$ and $C_3$ are the intersections of $A_1 A_3$, $B_1 B_3$ and $A_1 A_2$, $B_1 B_2$ respectively, then

$$C_2 = \lambda_3 A_3 - \lambda_1 A_1$$
$$C_3 = \lambda_1 A_1 - \lambda_2 A_2.$$

We see that $C_3 = -C_1 - C_2$. Thus $C_3$ is on $C_1 C_2$.

For the next theorem we need some preliminary definitions. A *hexagon* is a sequence of six points $A_1 A_2 A_3 A_4 A_5 A_6$, the vertices, of which no three are collinear, and of six lines $A_i A_{i+1}$ ($i = 1, \ldots, 6$; $A_7 = A_1$), the sides of the hexagon. $A_i$ and $A_{i+3}$ ($i = 1, 2, 3$) are opposite vertices; $A_i A_{i+1}$ and $A_{i+3} A_{i+4}$ are opposite sides. A line joining opposite vertices is a diagonal; the intersection of opposite sides is a diagonal point. A diagonal point $P$ *corresponds* to a diagonal $d$, if $d$ contains the two vertices which are not on a side through $P$.

**Theorem 1.4.6 (Pappos' theorem).** If two of the diagonal points of a hexagon are on their corresponding diagonals, then the same is true for the third diagonal point.

PROOF. Let $A_1 \ldots A_6$ be the vertices of the hexagon and $\alpha_1, \ldots, \alpha_6$ its sides, beginning with $A_1 A_2 = \alpha_1$, and ending with $A_6 A_1 = \alpha_6$. The intersections $\alpha_1 \cap \alpha_4$, $\alpha_2 \cap \alpha_5$, $\alpha_3 \cap \alpha_6$ are $C_1$, $C_2$, $C_3$ respectively. Assuming that $C_1$ is on $A_3 A_6$ and $C_2$ on $A_1 A_4$, we must prove that $C_3$ is on $A_2 A_5$.

Put $A_3 A_6 = \beta_1$, $A_1 A_4 = \beta_2$. We have $\alpha_4 = \nu\alpha_1 + \rho\beta_1$; as we may multiply $\alpha_1$ and $\beta_1$ each by a number, we may write

(1)                        $$\alpha_4 = \alpha_1 + \beta_1.$$

$\alpha_3 = \sigma\alpha_4 + \tau\beta_2$; multiplying $\alpha_3$ by $\sigma^{-1}$, $\beta_2$ by $\tau^{-1}$, we obtain

(2)                        $$\alpha_3 = \alpha_4 + \beta_2 = \alpha_1 + \beta_1 + \beta_2.$$

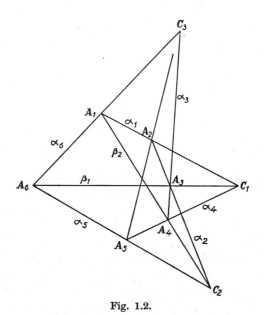

Fig. 1.2.

Similarly, keeping in mind that every newly introduced equation of a line may be multiplied by some number, we have

(3)                $$\alpha_2 = \alpha_3 + \kappa\beta_1 = \alpha_1 + \beta_2 + \lambda\beta_1 \quad (\lambda = \kappa + 1).$$

(4)     $\alpha_5 = \alpha_2 + \xi\beta_2 = \alpha_1 + \mu\beta_2 + \lambda\beta_1 \quad (\mu = \xi + 1).$

        $\alpha_6 = \alpha_5 + \theta\beta_1 = \alpha_1 + \mu\beta_2 + \eta\beta_1 \quad (\eta = \theta + \lambda).$

On the other hand, $\alpha_6 = \varepsilon\alpha_1 + \zeta\beta_2$.

The last two equations give $\eta\beta_1 = (\varepsilon - 1)\alpha_1 + (\zeta - \mu)\beta_2$. If $\eta$ were not 0, then $\beta_1$ would contain $A_1$, which is not the case; so $\eta = 0$.

(5)                   $\alpha_6 = \alpha_1 + \mu\beta_2.$

It is now easily verified from (1)—(5) that

$$(\lambda - 1)(\mu - 1)\alpha_1 + \mu\alpha_2 = \lambda(\mu - 1)\alpha_4 + \alpha_5 = \lambda\mu\alpha_3 + (1 - \mu)\alpha_6.$$

The line which has any of these expressions as the left-handside of its equation, passes through $A_2$, $A_5$ and $C_3$.

## Order relations in $APPG$.

Let $P$ and $Q$ be different points; a parametric representation of the line $PQ$ is $X = \lambda P + \mu Q$. The points of $PQ$, with the exception of $P$ and $Q$, can be divided into two classes $\Sigma_1$ and $\Sigma_2$, such that $\mu/\lambda > 0$ on $\Sigma_1$ and $\mu/\lambda < 0$ on $\Sigma_2$. $\Sigma_1$ and $\Sigma_2$ are called the *segments* $PQ$. If $R$ and $S$ are on different segments $PQ$, we say that $R$ and $S$ *separate* $P$ and $Q$.

**Theorem 1.4.7.** If $R$ and $S$ separate $P$ and $Q$, then $P$ and $Q$ separate $R$ and $S$.

Proof. Suppose $R = \lambda_1 P + \mu_1 Q$; $S = \lambda_2 P + \mu_2 Q$; then

$$(\lambda_1\mu_2 - \lambda_2\mu_1)P = \mu_2 R - \mu_1 S;$$

$$(\lambda_1\mu_2 - \lambda_2\mu_1)Q = -\lambda_2 R + \mu_2 S.$$

If $R$ and $S$ separate $P$ and $Q$, we have $\lambda_1\mu_1\lambda_2\mu_2 < 0$, which shows, by the last two equations, that $P$ and $Q$ separate $R$ and $S$.

If $\alpha$ and $\beta$ are different lines, then any line through their point of intersection $S$ has an equation of the form $\lambda\alpha + \mu\beta = 0$ (Th. 1.4.2). The lines of this pencil, with the exception of $\alpha$ and $\beta$, can be divided into two classes $\Phi_1$ and $\Phi_2$, such that $\mu/\lambda > 0$ for every line in $\Phi_1$ and $\mu/\lambda < 0$ for every line in $\Phi_2$.

$\Phi_1$ and $\Phi_2$ are *double-angles* in the pencil $S$. If two lines $\gamma$ and $\delta$ of the pencil are in different double-angles, we say that $\gamma$ and $\delta$ *separate* $\alpha$ and $\beta$.

**Theorem 1.4.8.** If $R$ and $S$ separate $P$ and $Q$, while $T$ is a point outside $PQ$, then $TR$ and $TS$ separate $TP$ and $TQ$.

Proof. We have $R = \lambda_1 P + \mu_1 Q$ and $S = \lambda_2 P + \mu_2 Q$ with $\lambda_1 \mu_1 \lambda_2 \mu_2 < 0$.

The equation of $TP$ is $\alpha = 0$, that of $TQ$ is $\beta = 0$, where

$$\alpha = \begin{vmatrix} x_0 & x_1 & x_2 \\ t_0 & t_1 & t_2 \\ p_0 & p_1 & p_2 \end{vmatrix}, \quad \beta = \begin{vmatrix} x_0 & x_1 & x_2 \\ t_0 & t_1 & t_2 \\ q_0 & q_1 & q_2 \end{vmatrix}.$$

The equation of $TR$ is

$$\begin{vmatrix} x_0 & x_1 & x_2 \\ t_0 & t_1 & t_2 \\ \lambda_1 p_0 + \mu_1 q_0 & \lambda_1 p_1 + \mu_1 q_1 & \lambda_1 p_2 + \mu_1 q_2 \end{vmatrix} = 0,$$

which is easily reduced to $\lambda_1 \alpha + \mu_1 \beta = 0$.

Similarly the equation of $TS$ is $\lambda_2 \alpha + \mu_2 \beta = 0$.

Then $\lambda_1 \mu_1 \lambda_2 \mu_2 < 0$ is exactly the condition that $TR$ and $TS$ separate $TP$ and $TQ$.

Corollary. The relation between separating pairs of points on a line is invariant under projection. In other words, if $PQ$ and $RS$ are separating pairs on a line $l$ and $P'$, $Q'$, $R'$, $S'$ are the projections of $P$, $Q$, $R$, $S$ on $m$ from a point $T$ outside $l$ and $m$, then $P'$ and $Q'$ separate $R'$ and $S'$.

The relations of separation between pairs of points of a line $l$ constitute what is called a *cyclical order* of $l$.

Here the development of $APPG$ will be discontinued; it is necessary to say a few words about analytic solid projective geometry ($ASPG$).


## § 1.5. Analytic solid projective geometry.

As most of what I have to say about it is similar to what has been said in § 1.4, I can be brief.

**Definition.** A point of $ASPG$ is a ratio of four numbers $(x_0, x_1, x_2, x_3)$; these numbers from a set of coordinates of the point.

**Definition.** A plane of *ASPG* is the set of points whose coordinates satisfy a linear equation

(1)    $a_0x_0 + a_1x_1 + a_2x_2 + a_3x_3 = 0$ (not every $a_i = 0$).

The plane is determined by the ratio $(a_0, a_1, a_2, a_3)$; the four coefficients form a set of coordinates of the plane.

**Definition.** A line is the intersection of two different planes. The following theorems follow from the theory of linear equations.

**Theorem 1.5.1.** If $P(p_i)$ and $Q(q_i)$ are different points of a line $l$, then a parameter representation of $l$ is

$$x_i = \lambda p_i + \mu q_i \quad (i = 1, 2, 3, 4).$$

**Theorem 1.5.2.** If $P(p_i)$, $Q(q_i)$, $R(r_i)$ are points in a plane $\alpha$, but not on a line, then a parameter representation of $\alpha$ is

$$x_i = \lambda p_i + \mu q_i + \nu r_i \quad (i = 1, 2, 3, 4).$$

**Theorem 1.5.3.** Given two different points, there is exactly one line which contains both of them.

**Theorem 1.5.4.** Given three points not on a line, there is exactly one plane which contains them.

**Theorem 1.5.5.** Given a line $l$ and a point $P$ not on $l$, there is exactly one plane which contains $P$ and $l$.

**Theorem 1.5.6.** Given a plane $\alpha$ and a line $l$ not in $\alpha$, there is exactly one point contained in $\alpha$ and in $l$.

## § 1.6. Vector spaces over a division ring.

The theory of vector spaces over a field is well known. In chapter 5 I need the corresponding theory for an arbitrary division ring. This theory is hardly more involved than that for the commutative case.

Let us summarize the main definitions and theorems.

We must distinguish between left vector spaces and right vector spaces. In the following the theory of left vector spaces is considered.

A left vector space $\mathfrak{V}$ over the division ring $\mathfrak{F}$ is a set of elements called vectors, in which two operations are defined, namely addition of two vectors, $\mathbf{a}+\mathbf{b}$, and multiplication of a vector by an element of $\mathfrak{F}$, $p\mathbf{a}$. (In the case of right vector spaces the element of $\mathfrak{F}$ is written to the right of the vector, i.e. $\mathbf{a}p$).

With respect to addition, $\mathfrak{V}$ is an abelian group; the multiplication satisfies the rules

$$p\mathbf{a}+p\mathbf{b} = p(\mathbf{a}+\mathbf{b})$$
$$p\mathbf{a}+q\mathbf{a} = (p+q)\mathbf{a}$$
$$p(q\mathbf{a})=(pq)\mathbf{a}$$
$$1\mathbf{a} = \mathbf{a}.$$

We can deduce that $0\mathbf{a} = \mathbf{0}$; here $\mathbf{0}$ is the zero of the addition of vectors.

The elements of $\mathfrak{F}$ are also called scalars.

The vectors $\mathbf{a}^1, \ldots, \mathbf{a}^k$ are *independent* if $p_1\mathbf{a}^1+ \ldots +p_k\mathbf{a}^k = 0$ implies $p_1 = \ldots = p_k = 0$. If the maximal number of independent vectors in $\mathfrak{V}$ is a finite number $n$, then $\mathfrak{V}$ is *$n$-dimensional*.

Let $\mathbf{e}^1, \ldots, \mathbf{e}^n$ be $n$ independent vectors in an $n$-dimensional vector space $\mathfrak{V}$; then every vector $\mathbf{x}$ in $\mathfrak{V}$ can be written uniquely in the form

$$(1) \qquad \mathbf{x} = x_1\mathbf{e}^1+ \ldots +x_n\mathbf{e}^n.$$

$x_1, \ldots, x_n$ are the coordinates of $\mathbf{x}$ with respect to the *basis* $\mathbf{e}^1, \ldots, \mathbf{e}^n$.

Thus by the choice of a basis $\mathbf{e}^1, \ldots, \mathbf{e}^n$, a one-to-one mapping $\varphi$ of $\mathfrak{V}$ onto the set $\mathfrak{F}^n$ of sequences $(x_1, \ldots, x_n)$ is established. If we make $\mathfrak{F}^n$ into a vector space $R_n(\mathfrak{F})$ by the definitions

$$(x_1, \ldots, x_n)+(y_1, \ldots, y_n)=(x_1+y_1, \ldots, x_n+y_n),$$
$$p(x_1, \ldots, x_n)=(px_1, \ldots, px_n),$$

$\varphi$ becomes an isomorphism between $\mathfrak{V}$ and $R_n(\mathfrak{F})$.

If $\mathbf{a}^1, \ldots, \mathbf{a}^k$ are independent vectors in $\mathfrak{V}$, then the linear combinations of $\mathbf{a}^1, \ldots, \mathbf{a}^k$ (i.e. the vectors which can be written as $p_1\mathbf{a}^1+ \ldots +p_k\mathbf{a}^k$) form a $k$-dimensional subspace of $\mathfrak{V}$. Another way to define linear subspaces is by linear equations. The solutions of a system of $r$ independent right-linear equations $\sum_i x_i a_{ik} = 0$ ($k = 1, \ldots, r$) form a $(n-r)$-dimensional subspace of $\mathfrak{V}$.

Let $\mathbf{u}^1, \ldots, \mathbf{u}^n$ be independent vectors in $\mathfrak{V}$;

$$(2) \qquad \mathbf{u}^i = \sum_k u_k^i \mathbf{e}^k.$$

Any vector $\mathbf{x}$ can be written uniquely as

(3)
$$\mathbf{x} = x_1' \mathbf{u}^1 + \ldots + x_n' \mathbf{u}^n.$$

Substituting (2) in (3), we obtain

$$\mathbf{x} = \sum_i x_i' \sum_k u_k^i \, \mathbf{e}^k = \sum_k \sum_i x_i' u_k^i \, \mathbf{e}^k.$$

Comparing with (1), we find

(4)
$$x_k = \sum_i x_i' u_k^i. \qquad (k = 1, \ldots, n)$$

(4) is the coordinate transformation from the basis $\mathbf{e}^1, \ldots, \mathbf{e}^n$ to the basis $\mathbf{u}^1, \ldots, \mathbf{u}^n$.

The proofs of these theorems can be given exactly as in the commutative case, the only difference being that attention must be given to the order of the factors in a product.

# CHAPTER II

# INCIDENCE PROPOSITIONS IN THE PLANE

## § 2.1. Trivial axioms, duality.

**Definition.** *A plane projective geometry* is an axiomatic theory with the triple $\langle \Pi, \Lambda, I \rangle$ as its set of fundamental notions and **V1, V2, V3** (formulated below) as its axioms, possibly with additional axioms. $\Pi$ and $\Lambda$ are disjoint sets and $I$ is a symmetric relation between $\Pi$ and $\Lambda$ (that is, if $a\,I\,b$ then either $a \in \Pi$ and $b \in \Lambda$ or $a \in \Lambda$ and $b \in \Pi$; $a\,I\,b$ is equivalent to $b\,I\,a$). The elements of $\Pi$ are called *points* those of $\Lambda$ are called *lines*; $I$ is the *incidence relation*. $a\,I\,b$ is read "$a$ is incident with $b$".

REMARKS. By what is said in § 1.1, it is clear that by using the words "point" and "line" we do not refer to any connection with space-intuition. On the contrary, we have to avoid the danger that these words lead us to an injustifiable appeal to space-intuition just the same as for the use of figures. Where it seems desirable the words "point" and "line" can be replaced by "element of $\Pi$" and "element of $\Lambda$".

The expression "the point $P$ is incident with the line $l$" or "the line $l$ is incident with the point $P$" was used above instead of "$P$ lies on $l$" or "$l$ passes through $P$" or "$l$ contains $P$". However, for the sake of liveliness of style, we shall often use one of the latter expressions instead of the former.

In future points will be denoted by capitals and lines by lower case letters.

## Axioms.

**V1a.** Given two different points, there is at least one line with which both are incident.

**V1b.** Given two different points, there is at most one line with which both are incident.

**V2.**   Given two different lines, there is at least one point with which both are incident.

**V3.**   $\Pi$ contains at least four points such that no three of them are incident with one and the same line, and at least two of them are different.

The axiomatic theory described above will be called $\mathfrak{P}$(V1, V2, V3), or briefly $\mathfrak{P}$. More generally the axiomatic theory with axioms V1, V2, V3, A1, A2, ..., Ak is denoted by $\mathfrak{P}$(A1, A2, ..., Ak). A theorem which can be derived in $\mathfrak{P}$ (in other words, from V1, V2, V3 alone) will be called *trivial*.

REMARKS. An axiomatic theory is a set of theorems which can be derived from the axioms. In the theory the fundamental notions are left unspecified. If we choose special notions satisfying the axioms, we obtain a model for the theory. For instance, a model for $\mathfrak{P}$ consists of disjunct sets $\Pi_0$ and $\Lambda_0$ and a relation $I_0$, which satisfy V1, V2, V3. Such a model is called a *projective plane*. Two projective planes $\langle \Pi_0, \Lambda_0, I_0 \rangle$ and $\langle \Pi_1, \Lambda_1, I_1 \rangle$ are isomorphic if there is a one-to-one mapping $\pi$ of $\Pi_0$ onto $\Pi_1$ and a one-to-one mapping $\lambda$ of $\Lambda_0$ onto $\Lambda_1$ so that $P I l$ if and only if $\pi P I \lambda l$. The latter condition is often expressed by saying that $\pi$ and $\lambda$ preserve incidence relations.

Different models for $\mathfrak{P}$(A1, ..., Ak) will often be denoted by $\mathfrak{P}_0$(A1, ..., Ak), $\mathfrak{P}_1$(A1, ..., Ak), etc.

The unique line which, according to V1, is incident with two different points $A$, $B$, will be designated as usual by $AB$.

The proposition which corresponds to V1b in the same way as V2 corresponds to V1a, is not taken as an axiom because it can be proved from V1, V2, V3 (Th. 2.1.1).

It has been proved in section 1.4 that $APPG$ is a model for the axiom system, consisting of V1, V2, V3.

**Theorem 2.1.1.** In $\mathfrak{P}$: Given two different lines, there is at most one point with which both are incident.

The proof is left to the reader. Note that only axiom V1b is needed!

**Theorem 2.1.2.** In $\mathfrak{P}$: $\Lambda$ contains at least four lines such that no three of them are incident with one and the same point.

PROOF. By V3 we can find four points $A_1$, $A_2$, $A_3$, $A_4$, such that no three of them are incident with one and the same line.

These four points are different (why?)[1]. We shall show that the four lines $A_1A_2$, $A_1A_3$, $A_2A_4$, $A_3A_4$ satisfy the condition of theorem 2.1.2.

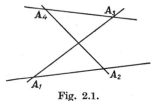

Fig. 2.1.

Suppose that $A_1A_2$, $A_3A_4$, $A_1A_3$ were incident with a point $P$. Then $A_1$ and $P$ are both incident with $A_1A_2$ as well as with $A_1A_3$, while $A_1A_2$ and $A_1A_3$ are different lines (why?). It follows from Th. 2.1.1. that $P = A_1$. In the same way we find that $P = A_3$, so $A_1 = A_3$, which is false. We have now shown that $A_1A_2$, $A_3A_4$ and $A_1A_3$ are not incident with a point. The proof for the other triples is analogous.

The reader must have been struck by a symmetry in these axioms and theorems which can be described as follows. If in V1a, V1b, V2 and V3 we interchange the words "point" and "line", we obtain V2, Th. 2.1.1, V1a and Th. 2.1.2 respectively.

Now let $\mathfrak{B}$ be a proof of a theorem $\Theta$ from V1, V2, V3. Let us interchange "point" and "line" in $\mathfrak{B}$ as well as in $\Theta$, obtaining $\mathfrak{B}'$ and $\Theta'$. Obviously $\mathfrak{B}'$ is a proof of $\Theta'$ from V2, Th. 2.1.1, V1a and Th. 2.1.2. But Th. 2.1.1 and Th. 2.1.2 can in their turn be derived from V1, V2, V3, so that $\Theta'$ can be derived from V1, V2, V3.

Thus we have proved:

**Theorem 2.1.3.** If in a theorem of $\mathfrak{P}$ we interchange the words "point" and "line", we obtain again a theorem of $\mathfrak{P}$.

Two theorems which change into each other if we interchange "point" and "line", are called *dual*. Theorem 2.1.3 expresses that the *duality principle* is valid for $\mathfrak{P}$.

---

[1] If we speak of two points, this does not mean that the points must be different. In the same way, if in arithmetic we speak of the sum of two numbers, these numbers can be equal.

REMARK. Th. 2.1.3 has a character different from other theorems, such as Th. 2.1.1 and Th. 2.1.2. It is not a theorem *in* $\mathfrak{P}$, but a theorem *on* $\mathfrak{P}$. It can also be expressed as follows: If $\langle \Pi_0, \Lambda_0, I_0 \rangle$ is a projective plane, then $\langle \Lambda_0, \Pi_0, I_0 \rangle$ is also a projective plane. These two planes are said to be *dual* to each other.

We shall now simplify our terminology by adapting it to common use.

**Definitions.** The *point of intersection* of two different lines is the point that is incident with both. The point of intersection of $l$ and $m$ is denoted by $l \cap m$.[1]) The *connecting line* of two different points is the line which is incident with both. The connecting line of $A$ and $B$ is denoted by $AB$. The points $A_i$ $(i = 1, \ldots, n)$ are *collinear* if there is a line with which each of them is incident. The lines $l_i$ $(i = 1, \ldots, n)$ are *concurrent* if there is a point with which each of them is incident.

A line is determined uniquely by the set of the points which are incident with it, and conversely. Therefore, no misunderstanding can arise if we identify a line with this set; accordingly we shall write "$P \in l$" (read: $P$ belongs to $l$) instead of "$P \, I \, l$"; for "not $P \, I \, l$" we write "$P \notin l$".

A *triangle* is a set of three different points $A_1$, $A_2$, $A_3$ and three lines $a_1$, $a_2$, $a_3$ such that $A_i \in a_k$ for $i \neq k$, but $A_i \notin a_i$ $(i, k = 1, 2, 3)$. The points $A_i$ are the *vertices*, the lines $a_i$ the *sides* of the triangle. The triangle is denoted by $A_1 A_2 A_3$. The vertex $A_i$ is *opposite* to the side $a_i$. The notation for a triangle is always so chosen, that a vertex and the opposite side are denoted by a capital and the corresponding lower case letter (with the same subscript).

The dual of a triangle is a *trilateral*. The description of this notion is left to the reader. He will notice that a triangle and a trilateral are the same figure, described in different ways.

---

[1]) If $P$ is the point of intersection, then, according to the current notation in set theory, $l \cap m = \{P\}$. I am sure that no confusion will be caused by the identification of $P$ and $\{P\}$.

## § 2.2. Desargues' proposition.

**Models.** Analytic plane projective geometry $APPG$ is a model of V1, V2, V3. Is this system of axioms complete for $APPG$? As it stands, this question has no precise meaning. We can now make it more precise: Can every theorem of $APPG$ which is formulated by means of the notions "point", "line" and "incident" alone, be derived in $\mathfrak{P}$ (V1, V2, V3)? The answer is in the negative. In fact, we shall construe a theorem of $APPG$, denoted (for a reason which will presently be explained) by $D_{11}$, and a model $M$ of $\mathfrak{P}$ (V1, V2, V3) in which $D_{11}$ is false. If $D_{11}$ were derivable in $\mathfrak{P}$ (V1, V2, V3), it should be true in every model of $\mathfrak{P}$, in particular in $M$.

$D_{11}$ is Desargues' proposition.

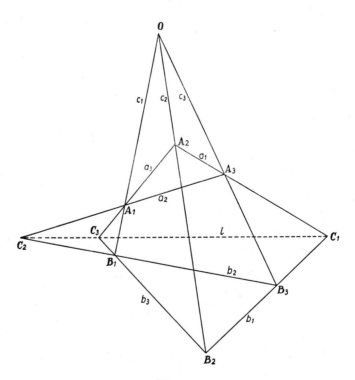

Fig. 2.2.

**Desargues' Proposition** $(D_{11})$. Let two triangles $A_1A_2A_3$ and $B_1B_2B_3$ be given. $A_i$ and $B_i$ are *corresponding* vertices; $a_i$ and $b_i$ are *corresponding* sides. If every two corresponding vertices as well as every two corresponding sides are different and the lines connecting corresponding vertices are incident with a point $O$, then the corresponding sides intersect in three collinear points.

$D_{11}$ is a theorem in $APPG$ (Th. 1.4.5).

REMARK. Note that we speak here of Desargues' proposition, not of Desargues' theorem. The reason was explained in section 1.1.3.

The model $M$ is defined as follows: There are three kinds of $M$-points, namely proper $M$-points, improper $M$-points and an extra $M$-point $Y$. A proper $M$-point is a pair of real numbers $(x, y)$. An improper $M$-point is a real number $(p)$.

There are also three kinds of $M$-lines. An $M$-line of the first kind is a pair of real numbers $[m, n]$. An $M$-line of the second kind is a real number $[c]$. There is one $M$-line $\omega$ of the third kind.

REMARK. The notation is so chosen, that numbers in ( ) indicate points; numbers in [ ] indicate lines.

The incidence relation $I_M$ is defined thus (read "if and only if" for "↔"):

$$(x, y) \, I_M \, [m, n] \leftrightarrow y = mx+n \text{ for } m \leq 0,$$
$$y = mx+n \text{ for } m > 0, \ x \leq 0,$$
$$y = 2mx+n \text{ for } m > 0, \ x > 0.$$
$$(p) \quad I_M \, [m, n] \leftrightarrow p = m.$$
$$(x, y) \, I_M \, [c] \quad \leftrightarrow x = c.$$
$$Y \quad I_M \, [c] \qquad \text{for every } c.$$
$$(p) \quad I_M \, \omega \qquad \text{for every } p.$$
$$Y \quad I_M \, \omega.$$

No incidences occur except those listed above.

$M$ is a model of $\mathfrak{P}$.

**Exercise.** The reader should verify this by examining all the cases for every axiom.

The verification is made easier by the following more intuitive but less precise description of the model $M$. In the Cartesian plane, completed by points at infinity, consider as $M$-lines the following sets:

$1^{st}$ kind, the Euclidean lines with non-positive slope, and the Euclidean lines with positive slope $m$, with the part at the right of the axis of $Y$ hinged up until the slope has become $2m$.

$2^{nd}$ kind, the lines parallel to the axis of $Y$.

$3^{rd}$ kind, the line at infinity.

To show that $D_{11}$ is not valid in $M$, we consider the triangles $A_1A_2A_3$ and $B_1B_2B_3$ with $A_1 = (0, 0)$, $A_2 = (-1, 1)$, $A_3 = (0, 2,)$ $B_1 = (2, 0)$, $B_2 = (1, 1)$, $B_3 = (2, 2)$. The $M$-lines $A_1B_1$, $A_2B_2$

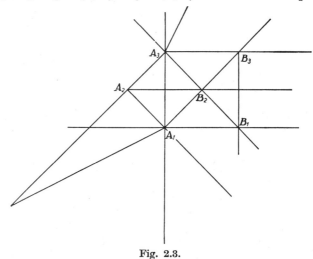

Fig. 2.3.

and $A_3B_3$ are incident with the $M$-point $(0)$. The three points of intersection of the corresponding sides are not collinear.

**Exercise.** The reader should verify this.

We have now proved:

**Theorem 2.2.1.** $D_{11}$ is independent of V1, V2, V3.

We can introduce $D_{11}$ as a new axiom and form $\mathfrak{P}$(V1, V2, V3, $D_{11}$), briefly denoted by $\mathfrak{P}(D_{11})$. But we shall first give Desargues' proposition a more general form which is more convenient in the applications.

**Generalized Desargues' Proposition.** $(D_{11}*)$ Let there be given 7 points $A_1 A_2 A_3 B_1 B_2 B_3 O$ and 9 lines $a_1 a_2 a_3 b_1 b_2 b_3 c_1 c_2 c_3$ such that none of the triples $A_1 A_2 A_3$ and $B_1 B_2 B_3$ consists of three coinciding points and that the following incidences take place:

$$A_1 \epsilon a_2, \ a_3, \ c_1; \ A_2 \epsilon a_3, \ a_1, \ c_2; \ A_3 \epsilon a_1, \ a_2, \ c_3;$$
$$B_1 \epsilon b_2, \ b_3, \ c_1; \ B_2 \epsilon b_3, \ b_1, \ c_2; \ B_3 \epsilon b_1, \ b_2, \ c_3;$$
$$O \epsilon c_1, \ c_2, \ c_3.$$

Then there exist three points $C_1, C_2, C_3$ and a line $l$ (not in every case unique) such that $C_1 \epsilon a_1, \ b_1, \ l; \ C_2 \epsilon a_2, \ b_2, \ l;$ $C_3 \epsilon a_3, \ b_3, \ l$.

REMARK. If in future we say: "apply the Generalized Desargues' Theorem to the points $PQR|STU|V$ and the lines $pqr|stu|vwx$" it is always tacitly understood that the order of the points and lines is the same as above, e.g. $P$ has the role of $A_1$, $Q$ of $A_2$, $V$ of $O$, $p$ of $a_1$, etc. The vertical lines serve to facilitate the reading.

**Theorem 2.2.2.** In $\mathfrak{P}(D_{11})$ the Generalized Desargues' Proposition is a theorem.

PROOF. We verify that the assertion is trivial in every case where the conditions of $D_{11}$ are not satisfied.

*Case* I. $a_1 = b_1$. Take $C_3$ such that $C_3 \epsilon a_3, b_3$ and $C_2$ such that $C_2 \epsilon a_2, b_2$; then $l$ such that $C_2, C_3 \epsilon l$ and $C_1$ such that $C_1 \epsilon a_1, l$.

Further on we may suppose that $a_i \neq b_i$; $C_i = a_i \cap b_i$ ($i = 1, 2, 3$).

*Case* II. $A_1 = B_1$. Then $C_2 = C_3 = A_1$. Take $l$ such that $C_1, A_1 \epsilon l$,

Further on we may suppose that $A_i \neq B_i$ ($i = 1, 2, 3$).

*Case* III. $A_1, A_2, A_3$ are different points on a line $l$. Take $C_i = b_i \cap l$.

*Case* IV. $A_1 = A_2 \neq A_3$. Then $a_1 = a_2$ and $C_1, C_2 \epsilon a_1$. If $C_3 \epsilon a_1$ we can take $l = a_1$. If $C_3 \notin a_1$, we consider two cases.

*Case* Va. $A_1 = A_2 \neq A_3$, $C_3 \notin a_1$, $B_1 \neq B_2$. From $A_1 C_3 = a_3$ and $a_3 \neq b_3$ we find $A_1 \notin b_3$, so $c_1 \neq c_2$. $O = c_1 \cap c_2 = A_1 \cdot c_3 = OA_3 = a_1$, so $B_3 \epsilon a_1$, so $B_3 = C_1$. Similarly $B_3 = C_2$. Take $l = C_1 C_3$.

*Case* Vb. Like Va, but $B_1 = B_2$. Then $b_1 = b_2$, so $C_1 = C_2$. Take for $l$ a line through $C_1$ and $C_3$.

Now the proof is complete.

**Exercise.** Show by a counterexample that the assertion in $D_{11}$ need not be true if $A_1 = A_2 = A_3$.

**Dual of Desargues' Proposition** $(dD_{11})$. Let two trilaterals $a_1a_2a_3$ and $b_1b_2b_3$ be given, such that corresponding sides as well as corresponding vertices are different. If corresponding sides intersect in points which are incident with a line $l$, then the lines connecting corresponding vertices are incident with a point $O$.

It is clear that $dD_{11}$ is also a converse of $D_{11}$.

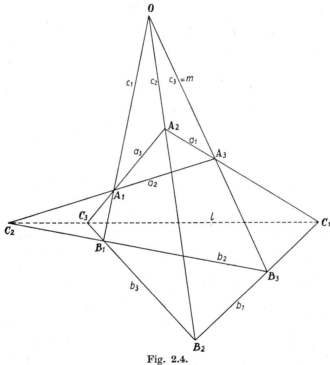

Fig. 2.4.

**Theorem 2.2.3.** $dD_{11}$ is a theorem in $\mathfrak{P}(D_{11})$.

**Proof.** Let $C_i$ be the point of intersection of $a_i$ and $b_i$; let $A_iB_i = c_i$ $(i = 1, 2, 3)$. We apply the Generalized Desargues' Theorem to the points $A_1B_1C_2|A_2B_2C_1|C_3$ and the lines $b_2a_2c_1|b_1a_1c_2|a_3b_3l$. There exist points $Q$, $P$, $O$ and a line $m$ such that $Q \in b_2$, $b_1$, $m$; $P \in a_2$, $a_1$, $m$; $O \in c_1$, $c_2$, $m$. It follows that $Q = B_3$, $P = A_3$, $m = c_3$, so that the theorem is proved.

From Th. 2.1.3 and Th. 2.2.3 we deduce immediately;

**Theorem 2.2.4.** The duality principle applies to $\mathfrak{P}(D_{11})$.

The figure of $D_{11}$ (fig. 2.4) is a configuration, consisting of 10 points and 10 lines; each of the points is incident with three of the lines and each of the lines is incident with three of the points. To each configuration-point $P$ there are three configuration-points which are not with $P$ on a configuration-line; these three points are on a line which is called *associated* with $P$.

For the application of $D_{11}$, we fix the following standard notation: Triangles $A_1A_2A_3$ and $B_1B_2B_3$; $A_2A_3 = a_1$, etc.;

$$A_iB_i = c_i;\ c_1 \cap c_2 \cap c_3 = O;\ a_i \cap b_i = C_i;\ C_1C_2 = l.$$

Then the following pairs are associated:
$A_i$ with $b_i$; $B_i$ with $a_i$; $C_i$ with $c_i$; $O$ with $l$.

The relation between a point and its associated line can be elucidated as follows. There is a one-to-one mapping of the 10 configuration-points onto the 10 combinations of two out of the numbers 1, 2, 3, 4, 5, such that the three points on a configuration-line correspond to the three combinations out of three of these numbers; accordingly, every configuration-line corresponds to a combination of three out of 1, 2, 3, 4, 5. For instance, such a mapping is given by the following table, where a point and its associated line are placed beside each other.

| | |
|---|---|
| $A_1$—14 | $b_1$—235 |
| $A_2$—24 | $b_2$—135 |
| $A_3$—34 | $b_3$—125 |
| $B_1$—15 | $a_1$—234 |
| $B_2$—25 | $a_2$—134 |
| $B_3$—35 | $a_3$—124 |
| $C_1$—23 | $c_1$—145 |
| $C_2$—13 | $c_2$—245 |
| $C_3$—12 | $c_3$—345 |
| $O$—45 | $l$—123 |

It is clear that the point $(pq)$ and the line $(rst)$ are associated if the numbers $p$, $q$, $r$, $s$, $t$ are all different. We shall acquire a deeper insight into this mapping by the use of solid geometry (see the remark after Th. 4.1.7).

Special cases of $D_{11}$ arise if we require that one or more configuration-points are incident with their associated lines. If this is the case for one point, this point can be either $O$, or one of the points $A_i$, $B_i$ or a point $C_i$.

**Exercise.** Verify that the assertion in $D_{11}$ becomes trivial if an extra incidence between a configuration-point and a non-associated configuration-line is postulated.

We shall treat in detail the case where $A_1 \epsilon b_1$.

**Small Desargues' Proposition** ($D_{10}$). Let two triangles $A_1A_2A_3$ and $B_1B_2B_3$ be given, such that corresponding vertices as well as corresponding sides are different, and that $A_1 \epsilon b_1$.

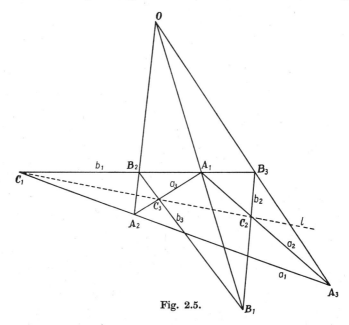

Fig. 2.5.

Let $a_i$ and $b_i$ intersect in $C_i$ ($i = 1, 2, 3$). If the lines connecting corresponding vertices are incident with a point $O$, then $C_1$, $C_2$, $C_3$ are collinear.

**Remark.** In $APPG$ the figure of $D_{11}$ depends upon 11 parameters; for instance, choose arbitrary points $O$, $A_1$, $A_2$, $A_3$ ($4 \times 2$ parameters) and $B_i$ on $OA_i$ ($3 \times 1$ parameter). The figure of the Small Desargues' Proposition depends upon 10 parameters. That is why Desargues' Proposition is called $D_{11}$ and the Small Desargues' Proposition $D_{10}$.

Of course, $D_{10}$ is a consequence of $D_{11}$. Later we shall see that $D_{11}$ is independent of V1, V2, V3, $D_{10}$.

$D_{10}$ can be generalized in the same way as $D_{11}$.

### Generalized Small Desargues' Proposition ($D_{10}^*$).

Let there be given 7 points $A_1 A_2 A_3 | B_1 B_2 B_3 | O$ and 9 lines $a_1 a_2 a_3 | b_1 b_2 b_3 | c_1 c_2 c_3$ such that none of the triples $A_1 A_2 A_3$ and $B_1 B_2 B_3$ consists of three coinciding points and that the incidences, postulated in $D_{11}^*$, take place, with in addition $A_1 \epsilon b_1$, then there exist points $C_1$, $C_2$, $C_3$ and a line $l$ such that $C_1 \epsilon a_1$, $b_1$, $l$; $C_2 \epsilon a_2$, $b_2$, $l$; $C_3 \epsilon a_3$, $b_2$, $l$.

**Theorem 2.2.5.** In $\mathfrak{P}(D_{10})$, $D_{10}^*$ is a theorem.

PROOF. In the proof of Th. 2.2.2 it has been verified that the assertion is trivial in every case where the assumptions of $D_{10}$ are not fulfilled.

The *dual of* $D_{10}$ (denoted by $dD_{10}$) can be formulated as follows:

Let two trilaterals $a_1 a_2 a_3$ and $b_1 b_2 b_3$ be given, such that corresponding sides as well as corresponding vertices are different, and that $b_1$ is incident with $A_1$. If corresponding sides intersect in points which are incident with a line $l$, then the lines connecting corresponding vertices are incident with a point $O$.

**Theorem 2.2.6.** $dD_{10}$ is a theorem in $\mathfrak{P}(D_{10})$.

PROOF. Like that of Th. 2.2.3, using $D_{10}^*$ instead of $D_{11}^*$.

It follows from Th. 2.1.3 and Th. 2.2.6:

**Theorem 2.2.7.** The duality principle applies to $\mathfrak{P}(D_{10})$.

By $D_{10}^{I}$ we shall denote Desargues' proposition with the extra hypothesis $O \epsilon l$, by $D_{10}^{II}$ that with the extra hypothesis $C_1 \epsilon c_1$.

**Proposition $D_{10}^{I}$.** Let $A_1 A_2 A_3$ and $B_1 B_2 B_3$ be triangles such that corresponding vertices as well as corresponding sides are different. Denote $A_i B_i$ by $c_i$, $a_i \cap b_i$ by $C_i$, and $C_1 C_3$ by $l$. If $c_1$, $c_2$, $c_3$ are incident with a point $O$, which lies on $l$, then $C_2$ lies on $l$.

**Theorem 2.2.8.** In $\mathfrak{P}(D_{10})$, $D_{10}^{I}$ is a theorem.

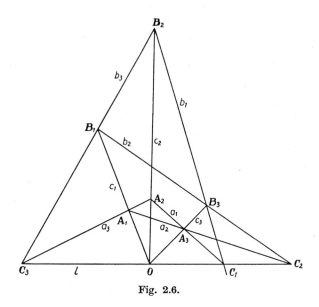

Fig. 2.6.

PROOF. Apply $D_{10}^*$ to the points and lines

$$OA_1A_3 \mid B_2C_3C_1 \mid A_2 \mid a_2c_3c_1 \mid l\,b_1b_3 \mid c_2a_3a_1, \text{ where } l = C_1C_3.$$

We find $P_1$, $P_2$, $P_3$, $m$, such that

$$P_1 \,\epsilon\, a_2,\ l,\ m;\quad P_2 \,\epsilon\, c_3,\ b_1,\ m;\quad P_3 \,\epsilon\, c_1,\ b_3,\ m.$$

In view of the remark after Th. 2.2.2 we may suppose that $b_1 \neq c_3$ and $b_3 \neq c_1$. Then we infer successively that $P_2 = B_3$, $P_3 = B_1$, $m = b_2$, $P_1 \,\epsilon\, a_2$, $b_2$, $P_1 = C_2$, $C_2 \,\epsilon\, l$.

The formulation of $D_{10}^{II}$ gives no difficulty.

**Theorem 2.2.9.** In $\mathfrak{P}(D_{10})$, $D_{10}^{II}$ is a theorem.

PROOF. Apply $D_{10}^*$ to the points and lines

$$C_1A_3C_2 \mid B_2OB_1 \mid B_3 \mid a_2la_1 \mid c_1b_3c_2 \mid b_1c_3b_2, \text{ where } l = C_1C_2.$$

By Th. 2.2.7, the duals of $D_{10}^I$ and $D_{10}^{II}$ are also theorems in $\mathfrak{P}(D_{10})$.

The converses of Th. 2.2.8 and Th. 2.2.9 also hold.

**Theorem 2.2.10.** In $\mathfrak{P}(D_{10}^I)$, $D_{10}$ is a theorem.

For the proof, we first have to extend $D_{10}^I$ to $D_{10}^{I*}$.

This is a little less easy than for $D_{10}$. It can be done as follows.

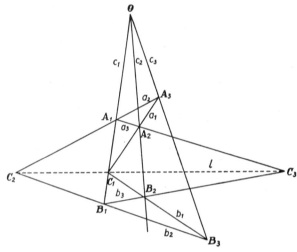

Fig. 2.7.

$D_{10}^{\text{I}}*$. Let points $A_1 A_2 A_3 B_1 B_2 B_3 O C_2 C_3$ and lines $a_1 a_2 a_3 b_1 b_2 b_3$ $c_1 c_2 c_3$ be given such that

| | | |
|---|---|---|
| $A_1 \epsilon a_2,\ a_3,\ c_1$ | $B_1 \epsilon b_2,\ b_3,\ c_1$ | $O \ \epsilon c_1,\ c_2,\ c_3,\ l$ |
| $A_2 \epsilon a_1,\ a_3,\ c_2$ | $B_2 \epsilon b_1,\ b_3,\ c_2$ | $C_2 \epsilon a_2,\ b_2,\ l$ |
| $A_3 \epsilon a_1,\ a_2,\ c_3$ | $B_3 \epsilon b_1,\ b_2,\ c_3$ | $C_3 \epsilon a_3,\ b_3,\ l;$ |

then there exist points $P_1$, $P_2$, $P_3$ and a line $m$, such that

$$P_1 \epsilon a_1,\ b_1,\ m,\ \ P_2 \epsilon a_2,\ b_2,\ m,\ \ P_3 \epsilon a_3,\ b_3,\ m.$$

We must prove that $D_{10}^{\text{I}}*$ is trivial in every case where the conditions of $D_{10}^{\text{I}}$ are not fulfilled.

This is done by simply repeating the proof of $D_{11}^*$.

Now $D_{10}$ follows by applying $D_{10}^{\text{I}}*$ to

$$O A_2 A_3 | B_1 C_3 C_2 | A_1 | a_1 c_3 c_2 | l b_2 b_3 | c_1 a_3 a_2 |, \text{ with } l = C_2 C_3,$$

the roles of $C_2$, $C_3$, $l$ being taken over by $B_3$, $B_2$, $b_1$ respectively. We find $P_1$, $P_2$, $P_3$, $m$, such that

$$P_1 \epsilon a_1,\ l,\ m;\ \ P_2 \epsilon c_3,\ b_2,\ m;\ \ P_3 \epsilon c_2,\ b_3,\ m.$$

It follows that $P_2 = B_3$, $P_3 = B_2$, $m = b_1$, $P_1 \epsilon a_1$, $b_1$, $P_1 = C_1$, $C_1 \epsilon l$.

Similarly, it can be proved that

**Theorem 2.2.11.** $D_{10}$ is a theorem in $\mathfrak{P}(D_{10}^{\text{II}})$.

Summarizing Theorems 2.2.8, 2.2.9, 2.2.10 and 2.2.11, we have

**Theorem 2.2.12.** In $\mathfrak{P}$, the propositions $D_{10}$, $D_{10}^{\mathrm{I}}$, $D_{10}^{\mathrm{II}}$ and their duals are all equivalent.

It was proved in Th. 2.2.1 that $D_{11}$ is independent of $V_1$, $V_2$, $V_3$. However, the triangles $A_1A_2A_3$ and $B_1B_2B_3$ in the proof of that theorem were such that $A_1 \in B_2B_3$; so $D_{10}$ does not hold in the model $M$. Therefore the following theorem, which is stronger than Th. 2.2.1, holds.

**Theorem 2.2.13.** $D_{10}$ is independent of $V_1$, $V_2$, $V_3$.

Let us now turn our attention to Desargues' proposition with two extra incidences in the hypothesis. If two points, which are not connected by a configuration-line, are on their associated lines (e.g. $A_1 \in b_1$, $C_1 \in c_1$), the assertion becomes trivial. Therefore only the following cases must be considered:

$$
\begin{array}{lll}
D_9 & A_1 \in b_1, & A_2 \in b_2. \\
D_9^{\mathrm{I}} & A_1 \in b_1, & B_1 \in a_1. \\
D_9^{\mathrm{II}} & A_1 \in b_1, & O \in l. \\
D_9^{\mathrm{III}} & A_1 \in b_1, & C_2 \in c_2. \\
D_9^{\mathrm{IV}} & C_1 \in c_1, & C_2 \in c_2.
\end{array}
$$

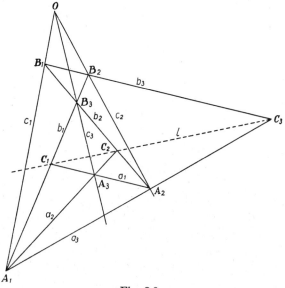

Fig. 2.8.

**Proposition** $D_9$. Let $A_1 A_2 A_3$ and $B_1 B_2 B_3$ be triangles such that corresponding vertices as well as corresponding sides are different. Denote $A_i B_i$ by $c_i$, $a_i \cap b_i$ by $C_i$. If $c_1$, $c_2$, $c_3$ are incident with a point $O$, and $A_1 \in b_1$ and $A_2 \in b_2$, then $C_1$, $C_2$, $C_3$ are collinear.

**Generalized proposition** $D_9$ (denoted by $D_9{}^*$). As $D_{11}{}^*$, with the extra postulated incidences $A_1 \in b_1$, $A_2 \in b_2$.

**Theorem 2.2.14.** In $\mathfrak{P}(D_9)$, $D_9{}^*$ is a theorem.

PROOF. See the proof of Th. 2.2.2.

**Theorem 2.2.15.** In $\mathfrak{P}(D_9)$, the dual of $D_9$, denoted by $dD_9$, is a theorem.

PROOF. $dD_9$ can be formulated as follows.

Let the trilaterals $a_1 a_2 a_3$ and $b_1 b_2 b_3$ be given, such that corresponding sides as well as corresponding vertices are different. Let $A_i B_i$ be denoted by $c_i$, $a_i \cap b_i$ by $C_i$. If $C_1$, $C_2$, $C_3$ are on a line $l$, while $A_1 \in b_1$, $A_2 \in b_2$, then $c_1$, $c_2$, $c_3$ pass through a point $O$.

Let us denote $c_1 \cap c_2$ by $O$; we prove that $O \in c_3$.

By the remark following Th. 2.2.2 we may suppose that $B_3 \neq O$, $O \notin a_2$, $C_2 \neq C_3$, $b_1 \neq l$, $A_2 \neq C_1$. Let us provisionally denote $OB_3$ by $c_3'$, $c_3' \cap a_2$ by $A_3'$, $A_2 A_3'$ by $a_1'$. Apply $D_9{}^*$ to

$$A_1 A_2 A_3' \mid B_1 B_2 B_3 \mid O \mid a_1' a_2 a_3 \mid b_1 b_2 b_3 \mid c_1 c_2 c_3'.$$

We find $P_1$, $P_2$, $P_3$, $m$, such that

$$P_1 \in a_1', \ b_1, \ m; \quad P_2 \in a_2, \ b_2, \ m; \quad P_3 \in a_3, \ b_3, \ m.$$

It follows successively that $P_2 = C_2$; $P_3 = C_3$; $m = l$; $P_1 \in b_1$, $l$; $P_1 = C_1$; $C_1 \in a_1'$; $a_1' = A_2 C_1 = a_1$; $A_3' = a_1 \cap a_2 = A_3$; $c_3' = c_3$; $O \in c_3$.

**Theorem 2.2.16.** In $\mathfrak{P}(D_9)$, the duality principle holds.

PROOF. From Th. 2.1.3 and Th. 2.2.15.

**Theorem 2.2.17.** In $\mathfrak{P}(D_9)$, $D_9^{\mathrm{I}}$, $D_9^{\mathrm{II}}$, $D_9^{\mathrm{III}}$ and $D_9^{\mathrm{IV}}$, and their duals, are theorems.

PROOF. a). For $D_9^{\mathrm{I}}$. Instead, we shall prove the dual $dD_9^{\mathrm{I}}$, which is obtained from $dD_9$ by changing the condition $A_2 \in b_2$ into $B_1 \in a_1$.

Apply $D_9{}^*$ to

$$A_1 B_1 C_2 \mid A_2 B_2 C_1 \mid C_3 \mid b_2 a_2 c_1 \mid b_1 a_1 c_2 \mid a_3 b_3 \, l.$$

We obtain $P_1$, $P_2$, $P_3$, $m$, such that

$$P_1 \in b_2,\ b_1,\ m;\quad P_2 \in a_2,\ a_1,\ m;\quad P_3 \in c_1,\ c_2,\ m.$$

It follows that $P_1 = B_3$; $P_2 = A_3$; $m = c_3$; $P_3 \in c_1$, $c_2$, $c_3$. By Th. 2.2.16 we can infer $D_9^I$ from $dD_9^I$.

b). For $D_9^{II}$. The proof is identical with that for $D_{10}^I$ (Th. 2.2.8).

c). For $D_9^{III}$. If we take for the extra incidence relations $A_3 \in b_3$, $C_1 \in c_1$, we may repeat the proof of $D_{10}^{II}$ (Th. 2.2.9).

d). For $D_9^{IV}$. The proof of $D_{10}^{II}$ can be used after interchanging the second and third vertices of each of the triangles: Apply $D_9{}^*$ to

$$C_1 C_2 A_3 \mid B_2 B_1 0 \mid B_3 \mid a_2 a_1 l \mid c_1 c_2 b_3 \mid b_1 b_2 c_3, \text{ where } l = C_1 C_2.$$

In fact, $D_9$, $D_9^I$, $D_9^{II}$, $D_9^{III}$, $D_9^{IV}$ and their duals, are all equivalent. We shall prove that $D_9^I$ implies $D_9$; the proofs of the other cases are left to the reader.

**Theorem 2.2.18.** In $\mathfrak{P}(D_9^I)$, $D_9$ is a theorem.

PROOF. First generalize $D_9^I$ to $D_9^I{}^*$ in the obvious way. We wish to prove $dD_9$ as formulated in the proof of Th. 2.2.15. Apply $D_9^I{}^*$ to

$$A_1 B_1 C_2 \mid A_2 B_2 C_1 \mid C_3 \mid b_2 a_2 c_1 \mid b_1 a_1 c_2 \mid a_3 b_3 l.$$

We obtain points $P_1$, $P_2$, $P_3$ and a line $m$, such that

$$P_1 \in b_2,\ b_1,\ m;\quad P_2 \in a_2,\ a_1,\ m;\quad P_3 \in c_1,\ c_2,\ m.$$

It follows that $P_1 = B_3$, $P_2 = a_3$; $m = c_3$; $P_3 \in c_1$, $c_2$, $c_3$. This proves $dD_9$.

By the dual of Th. 2.2.15, $D_9$ holds in $\mathfrak{P}(dD_9)$.

**Definition.** If $P$ is a fixed point and $s$ a given line, then $D_{11}(P, s)$ is $D_{11}$ with the additional conditions $0 = P$, $l = s$ (in the standard notation; see after Th. 2.2.4). Analogously, if $P \in s$, $D_{10}(P, s)$ will denote $D_{10}$ with the additional conditions $0 = P$, $l = s$, etc.

## § 2.3. Collineations.

**Definition.** A collineation is a one-to-one mapping of $\Pi$ onto itself, in which the image of every line is a line.

In other words: A collineation consists of a one-to-one transformation $\pi$ of $\Pi$ onto itself, and a one-to-one transformation $\lambda$ of $\Lambda$ onto itself, which preserve incidence relations (that is, such that $P I l$ entails $\pi P I \lambda l$). This is also expressed by saying that a collineation is an automorphism of the projective plane.

In $\mathfrak{P}$ the collineations form a group, the *collineation group* $\mathfrak{G}(\mathfrak{P})$. At present we are not able to decide whether $\mathfrak{G}(\mathfrak{P})$ contains any element different from the identity.

**Definition.** If every line through $C$ is invariant under the collineation $\mathfrak{C}$, then $C$ is called a *centre* of $\mathfrak{C}$. A collineation with a centre is called a *central collineation.*

The following theorem is easy to prove.

**Theorem 2.3.1.** In $\mathfrak{P}$:

I. A centre of a collineation $\mathfrak{C}$ is an invariant point of $\mathfrak{C}$.

II. A collineation with two different centres is the identity.

III. If in a central collineation $\mathfrak{C}$ with centre $C$ the line $l$, not through $C$, is invariant, then every point of $l$ is invariant.

IV. A collineation with two different lines of invariant points is the identity (dual of II).

**Theorem 2.3.2.** In $\mathfrak{P}$: Every central collineation which is not the identity has one and only one line of invariant points.

Proof. Let $\mathfrak{C}$ be a central collineation with centre $C$. We write $P'$ for $\mathfrak{C}P$, $l'$ for $\mathfrak{C}l$, etc. Choose $P$ such that $P \neq P'$; then $P \neq C$ and $P' \in PC$. If $a$ is a line such that $P \in a$, $C \notin a$, then $a \neq a'$, and $A = a \cap a'$ is an invariant point, for $A' = CA \cap a' = A$. As there are at least three lines through $P$ (which follows from V3) we find another invariant point $B$. $AB = d$ is an invariant line. If $C \notin d$, it is clear by Th. 2.3.1, III that every point of $d$ is invariant. If $C \in d$ we reason as follows. Let $X$ be any point on $d$, $X \neq C$, and let $l$ be a line through $X$ such that $l \neq l'$ (such a line exists by Th. 2.3.1, II). $l \cap l' = Y$. Suppose $Y \notin d$; then $YA$ and $YB$ are invariant lines not through $C$; it follows that every point of $YA$ as well as of $YB$ is invariant, which is impossible by Th. 2.3.1, IV. Thus $Y \in d$, $Y = X$, and $X$ is invariant. From

Th. 2.3.1, IV it follows immediately that there is only one line $d$ of invariant points.

**Definition.** The line of invariant points of a central collineation $\mathfrak{C}$ is called the *axis* of $\mathfrak{C}$. If the centre of $\mathfrak{C}$ lies on the axis, $\mathfrak{C}$ is a *special* central collineation.

**Theorem 2.3.3.** In $\mathfrak{P}$: Let there be given a point $O$, a line $l$ and points $P$, $P'$, such that $O$, $P$, $P'$ are collinear, $P \neq O$, $P' \neq O$, $P \notin l$, and $P' \notin l$. Then there exists at most one central collineation $\mathfrak{C}$ with centre $O$ and axis $l$ such that $\mathfrak{C}P = P'$.

PROOF. Let $A$ be any point outside $OP$. If a collineation $\mathfrak{C}$ as mentioned in the theorem exists, then $A' = \mathfrak{C}A$ can be constructed as follows. $PA \cap l = S$, $OA \cap P'S = A'$. If $B$ is a point on $OP$, then $B' = \mathfrak{C}B$ is constructed analogously from $A$ and $A'$. Thus the image by $\mathfrak{C}$ of every point is uniquely determined.

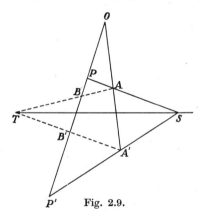

Fig. 2.9.

**Definition.** In this section we shall say that $P$, $P'$, $O$, $l$ are in *allowed position*, if $P \neq O$, $P' \neq O$, $P \notin l$, $P' \notin l$, $OP = OP'$. Throughout the book the central collineation with centre $O$ and axis $l$ which transforms $P$ into $P'$ will be denoted, if it exists, by $(PP'O^2l^2)$. Where this notation is used, it is tacitly understood that $P$, $P'$, $O$, $l$ are in allowed position.

Theorem 2.3.3 asserts, that $(PP'O^2l^2)$, if it exists, is uniquely determined. We shall now investigate under which conditions it exists.

**Definition.** If, for a given point $O$ and a given line $l$, the central collineation $(PP'O^2 l^2)$ exists for every choice of $P$, $P'$ such that $P$, $P'$ $O$, $l$ are in allowed position, then the plane is called $O$-$l$-*transitive*.

**Theorem 2.3.4.** In $\mathfrak{P}$: $D_{11}(O, l)$ implies that the plane is $O$-$l$-transitive.

Proof. Let $P$, $P'$ be points such that $P$, $P'$, $O$, $l$ are in allowed position. Let $A$ be any point outside $OP$. We repeat the construction in the proof of Th. 2.3.3. $PA \cap l = S$, $OA \cap P'S = A'$. This construction results in a transformation $\mathfrak{S}_{PP'}$ which is defined for every point outside $OP$. $A' = \mathfrak{S}_{PP'}(A)$.

The rest of the proof will be given in the form of three lemmas.

**Lemma 1.** If $Q \notin OP$, $Q \notin l$, $Q' = \mathfrak{S}_{PP'}(Q)$, $A \notin OP$ and $A \notin OQ$, then $\mathfrak{S}_{PP'}(A) = \mathfrak{S}_{QQ'}(A)$.

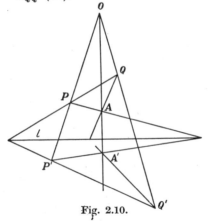

Fig. 2.10.

Proof. If $P = P'$ or $A \epsilon PQ$ or $A \epsilon l$, the result is obvious. So let us suppose that $P \neq P'$, $A \notin PQ$ and $A \notin l$. Denoting $\mathfrak{S}_{PP'}(A)$ by $A'$, we apply $D_{11}(O, l)$ to the triangles $PQA$ and $P'Q'A'$. Firstly, it follows that $QA$ intersects $Q'A'$ on $l$; and secondly, that $\mathfrak{S}_{QQ'}(A) = A'$.

**Lemma 2.** Let $Q$, $R$ be points outside $OP$ and $l$, and $A$ a point on $OP$, but $A \neq O$. $\mathfrak{S}_{PP'}(Q) = Q'$; $\mathfrak{S}_{PP'}(R) = R'$. Then $\mathfrak{S}_{QQ'}(A) = \mathfrak{S}_{RR'}(A)$.

Proof. First suppose that $OQ \neq OR$. By lemma 1 we have

$\mathfrak{S}_{QQ'}(R) = R'$; another application of lemma 1 gives the desired result. If $OQ = OR$, we choose $S$ outside $OP$, $OQ$ and $l$; then $\mathfrak{S}_{QQ'}(A) = \mathfrak{S}_{SS'}(A) = \mathfrak{S}_{RR'}(A)$.

We are now able to extend $\mathfrak{S}_{PP'}$ to points of $OP$: Let $Q$ be any point outside $OP$ and $l$; $\mathfrak{S}_{PP'}(Q) = Q'$; then for $A$ on $OP$ ($A \neq O$) we define $\mathfrak{S}_{PP'}(A) = \mathfrak{S}_{QQ'}(A)$. Lemma 2 shows that the point $A' = \mathfrak{S}_{PP'}(A)$ thus defined, does not depend upon the choice of $Q$. Moreover, we define $\mathfrak{S}_{PP'}(O) = O$.

**Lemma 3.** The mapping $\mathfrak{S}_{PP'}$, thus extended, is a central collineation.

PROOF. We must show that the image of a line $a$ is a line. This is clear if $O \,\epsilon\, a$ and if $a = l$. Suppose $O \,\xcancel{\epsilon}\, a$ and $a \neq l$; $a \cap l = S$. Choose $Q$ on $a$, $Q \,\xcancel{\epsilon}\, OP$, $Q \,\xcancel{\epsilon}\, l$. $\mathfrak{S}_{PP'}(Q) = Q'$. By lemmas 1 and 2 $\mathfrak{S}_{PP'} = \mathfrak{S}_{QQ'}$, but $\mathfrak{S}_{QQ'}(a) = SQ'$. This proves the lemma, and thereby Th. 2.3.4.

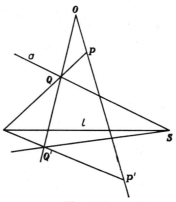

Fig. 2.11.

As a special case of Th. 2.3.4 we note:

**Theorem 2.3.5.** In $\mathfrak{P}$: $D_{10}^{1}(O, l)$, where $O \,\epsilon\, l$, implies that the plane is $O$-$l$-transitive.

As an immediate consequence of Th. 2.3.4 we have:

**Theorem 2.3.6.** In $\mathfrak{P}(D_{11})$: If $P$, $P'$, $O$, $l$ are in allowed position, then the central collineation $(PP'O^2l^2)$ exists.

Similarly, from Th. 2.3.5, together with Th. 2.2.8:

**Theorem 2.3.7.** In $\mathfrak{P}(D_{10})$: If $O \,\epsilon\, l$ and if $P$, $P'$, $O$, $l$ are in allowed position, then the special central collineation $(PP'O^2 l^2)$ exists.

The converses of Theorems 2.3.4—2.3.7 are also true:

**Theorem 2.3.8.** In $\mathfrak{P}$: If the plane is $O$-$l$-transitive, then $D_{11}(O, l)$ is valid.

**Theorem 2.3.9.** In $\mathfrak{P}$: If $O \in l$ and the plane is $O$-$l$-transitive, then $D_{10}^{\mathrm{I}}(O, l)$ is valid.

**Theorem 2.3.10.** In $\mathfrak{P}$: If every central collineation $(PP'O^2 l^2)$ exists, where $P$, $P'$, $O$, $l$ are in allowed position, then $D_{11}$ is valid.

**Theorem 2.3.11.** In $\mathfrak{P}$: If every special central collineation $(PP'O^2 l^2)$ exists, where $P$, $P'$, $O$, $l$ are in allowed position and $O \in l$, then $D_{10}^{\mathrm{I}}$ is valid.

PROOFS. It suffices to prove Th. 2.3.8. Let $A_1 A_2 A_3$ and $B_1 B_2 B_3$ be triangles such that corresponding vertices as well as corresponding sides are different, while $O \in A_i B_i$ $(i = 1, 2, 3)$. $a_i \cap b_i = C_i$ $(i = 1, 2, 3)$; $C_1 C_2 = l$. The central collineation $(A_1 B_1 O^2 l^2)$ transforms $a_3$ into $b_3$; thus $C_3 \in l$.

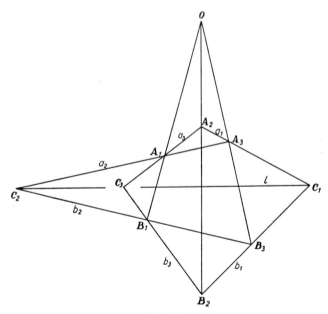

Fig. 2.12.

## § 2.4. The first quadrangle proposition, harmonic pairs.

**Definition.** A *complete quadrangle* is a set of 4 points $A_i$ ($i = 1, 2, 3, 4$), of which no three are collinear, and of six lines $l_{ik} = A_i A_k$ ($i \neq k$).

**First Quadrangle Proposition** ($Q_1$). Let $A_1 A_2 A_3 A_4$ and $B_1 B_2 B_3 B_4$ be two complete quadrangles and $l$ a line which contains none of the points $A_i$, $B_i$ ($i = 1, 2, 3, 4$). If $A_i A_j \cap l = B_i B_j \cap l = P_{ij}$ for $(ij) = (12), (13), (14), (23)$ and $(24)$, and if $A_3 A_4 \cap l = P_{34}$, then $B_3 B_4 \cap l = P_{34}$.

**Theorem 2.4.1.** (i). $Q_1$ can be derived in $\mathfrak{P}(D_{11})$.

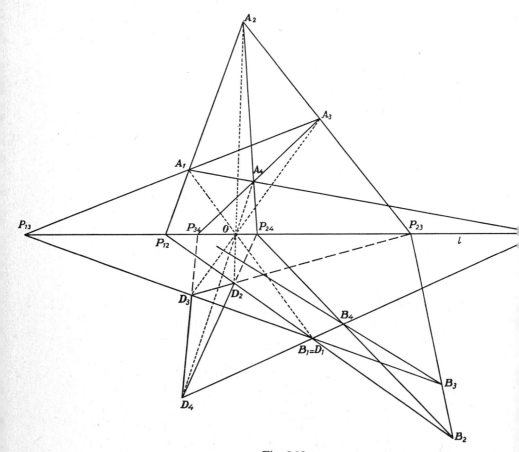

Fig. 2.13.

PROOF. Suppose that $A_1 \neq B_1$. Apply the special central collineation $\mathfrak{C}_1 = (A_1 B_1 O^2 l^2)$ with $O = A_1 B_1 \cap l$. $\mathfrak{C}_1 A_i = D_i$. By the properties of the central collineation, $D_1 D_2 D_3 D_4$ is a complete quadrangle; its vertices are outside $l$ and $D_i D_k \cap l = A_i A_k \cap l$; in particular $B_1 (=D_1)$, $B_2$ and $D_2$ are collinear. Now apply the central collineation $\mathfrak{C}_2 = (D_2 B_2 B_1^2 l^2)$; $\mathfrak{C}_2 D_i = B_i$ $(i = 1, 2, 3, 4)$. Therefore, $B_3 B_4 \cap l = D_3 D_4 \cap l = A_3 A_4 \cap l = P_{34}$.

**Exercise.** The reader should prove the theorem for $A_1 = B_1$, $A_2 \neq B_2$ and for $A_1 = B_1$, $A_2 = B_2$.

**Theorem 2.4.1.** (ii). Conversely, $D_{11}$ can be derived in $\mathfrak{P}(Q_1)$

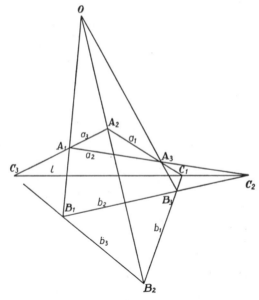

Fig. 2.14.

PROOF. Let the triangles $A_1 A_2 A_3$ and $B_1 B_2 B_3$ and the point $O$ be given such that $A_i \neq B_i$, $a_i \neq b_i$ and $O \in A_i B_i$ $(i = 1, 2, 3)$. $a_i \cap b_i = C_i$; $C_1 C_2 = l$. We may suppose that $O A_1 A_2 A_3$ and $O B_1 B_2 B_3$ are complete quadrangles, since otherwise $D_{11}$ is trivial (verify this). If $O \notin l$, the conditions of $Q_1$ are fulfilled and $a_3 \cap l = b_3 \cap l = C_3$. The case $O \in l$ must be treated separately. Apply $Q_1$ to the complete quadrangles $O A_1 A_2 A_3$ and $B_3 C_2 C_1 A_3$, together with the line $B_1 B_2$ (fig. 2.15). It follows that $B_1 B_2 \cap A_1 A_2 = B_1 B_2 \cap C_2 C_1 = C_3$.

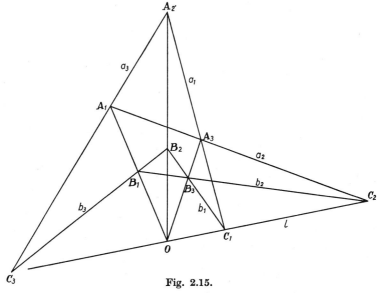

Fig. 2.15.

**Exercise.** The last part of the proof is not correct if $A_3 \in B_1 B_2$. The reader may prove this case by changing subscripts and letters. We summarize theorems 2.4.1(i) and 2.4.1(ii) in:

**Theorem 2.4.1.** $D_{11}$ and $Q_1$ are equivalent in $\mathfrak{P}$.

## Harmonic Pairs.

**Definition.** The points $Q$, $Q'$ are *harmonic* with $P$, $P'$ (notation: $Q$, $Q'$ harm. $P$, $P'$) if there exists a line $l$ and a complete quadrangle

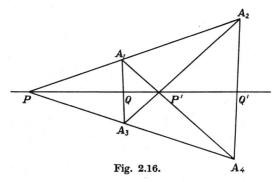

Fig. 2.16.

$A_1 A_2 A_3 A_4$ such that $P$, $P'$, $Q$, $Q' \,\epsilon\, l$, $P = A_1 A_2 \cap A_3 A_4$, $P' = A_1 A_4 \cap A_2 A_3$, $Q = l \cap A_1 A_3$, $Q' = l \cap A_2 A_4$. $Q'$ is called a *harmonic conjugate* of $Q$ with respect to $P$, $P'$.

If we speak of the harmonic conjugate of $Q$ with respect to $P$, $P'$, it will be tacitly understood that $P$, $P'$, $Q$ are different points on a line.

REMARK. It follows from the definition of a complete quadrangle, that if $Q$, $Q'$ are harmonic with $P$, $P'$, then $P \neq P'$ and $Q$, $Q'$ are different from $P$, $P'$; however, it does not follow that $Q \neq Q'$. We shall see (§ 6.2) that indeed the last inequality does not follow from the axioms.

**Exercise.** Verify that $Q$, $Q'$ harm. $P$, $P'$ entails $Q'$, $Q$ harm. $P$, $P'$ and $Q$, $Q'$ harm. $P'$, $P$.

As a special case of Th. 2.4.1(i) we have:

In $\mathfrak{P}(D_{11})$ the harmonic conjugate of $Q$ with respect to $P$, $P'$ is uniquely determined.

However, we shall prove the stronger:

**Theorem 2.4.2.** In $\mathfrak{P}(D_9)$: If $P$, $P'$, $Q$ are collinear and different, then the harmonic conjugate of $Q$ with respect to $P$, $P'$ is uniquely determined. In other words: If $A_1 A_2 A_3 A_4$ is a quadrangle as in the definition above, and if $B_1 B_2 B_3 B_4$ is a second quadrangle such that $B_1 B_2 \cap B_3 B_4 = P$, $B_1 B_4 \cap B_2 B_3 = P'$, $l \cap B_1 B_3 = Q$, then $l \cap B_2 B_4 = Q'$.

PROOF. First of all, we prove Th. 2.4.2 in two special cases, in which a vertex of the first quadrangle coincides with the corresponding vertex of the second.

**Case a).** The coinciding vertices are joined to $Q$, e.g. $A_1 = B_1$. $dD_9^I$ can be applied to the triangles $P' A_2 B_2$ and $P A_4 B_4$, since $P' \,\epsilon\, A_4 B_4$, $P \,\epsilon\, A_2 B_2$, and the points of intersection $A_2 B_2 \cap A_4 B_4 = A_1$, $A_2 P' \cap A_4 P = A_3$, $B_2 P' \cap B_4 P = B_3$ are collinear. It follows that $PP'$, $A_2 A_4$ and $B_2 B_4$ are concurrent; that is, $B_2 B_4$ passes through $Q'$. (Fig. 2.17).

**Case b).** The coinciding vertices are not joined to $Q$, e.g. $A_2 = B_2$. Now $D_9^I$ can be applied to the triangles $P' A_1 B_1$ and $P A_3 B_3$, for $P \,\epsilon\, A_1 B_1$, $P' \,\epsilon\, A_3 B_3$ and $A_1 A_3$, $B_1 B_3$, $PP'$ are concurrent. It follows that the points of intersection

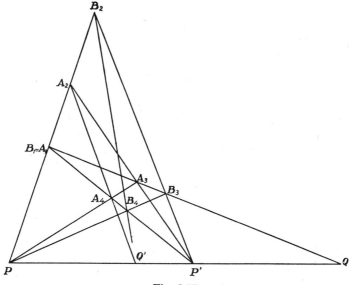

Fig. 2.17.

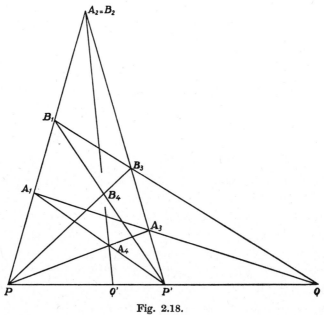

Fig. 2.18.

$A_1 B_1 \cap A_3 B_3 = A_2$, $A_1 P' \cap A_3 P = A_4$ and $B_1 P' \cap B_3 P = B_4$ are collinear; that is, $B_2 B_4 = A_2 A_4$, so $B_2 B_4$ passes through $Q'$. (Fig. 2.18).

**General case.**  Consider two intermediate quadrangles $R_1 R_2 R_3 R_4$ and $S_1 S_2 S_3 S_4$, where $R_1 = A_1 A_2 \cap B_1 B_3$, $R_2 = A_2$, $R_3 = A_2 A_3 \cap B_1 B_3$, $R_4 = PR_3 \cap P'R_1$, $S_1 = R_1$, $S_2 = A_1 A_2 \cap B_2 B_3$,

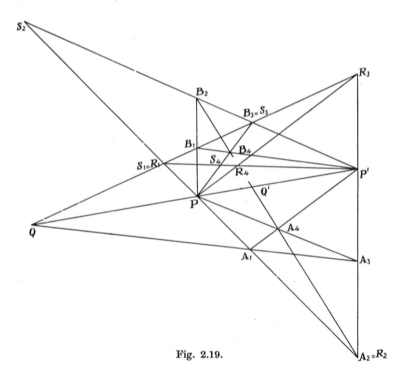

Fig. 2.19.

$S_3 = B_3$, $S_4 = PS_3 \cap P'S_1$. The quadrangles $A_1 A_2 A_3 A_4$ and $R_1 R_2 R_3 R_4$ are as in case b), so that $A_4$, $R_4$, $A_2$ are collinear; $R_2 R_4$ passes through $Q'$. $R_1 R_2 R_3 R_4$ and $S_1 S_2 S_3 S_4$ are as in case a), so that $S_2$, $S_4$, $Q'$ are collinear. $S_1 S_2 S_3 S_4$ and $B_1 B_2 B_3 B_4$ are again as in case a), so that $B_2$, $B_4$, $Q'$ are collinear.

REMARK. The figure of case b) shows that this special case of theorem 2.4.2 is equivalent to $D_9^{\mathrm{I}}$; thus $D_9^{\mathrm{I}}$ follows from the uniqueness of the harmonic conjugate. We have now:

**Theorem 2.4.3.** $D_9^I$, and consequently $D_9$, is equivalent in $\mathfrak{P}$ to the uniqueness of the harmonic conjugate.

**Theorem 2.4.4.** In $\mathfrak{P}(D_9)$: If $Q$, $Q'$ are harmonic with $P$, $P'$, and $Q \neq Q'$, then $P$, $P'$ are harmonic with $Q$, $Q'$.

PROOF. With the notation of fig. 2.20, put $B_2 = PA_4 \cap Q'A_1$, $B_4 = P'A_4 \cap Q'A_3$. $D_9$ applied to the triangles $A_1A_3Q'$ and $PP'A_4$ proves that $B_2B_4$ passes through $Q$. The quadrangle $A_3A_1B_2B_4$ shows that $P$, $P'$ are harmonic with $Q$, $Q'$.

Thus, in $\mathfrak{P}(D_9)$ we can speak of two harmonic pairs $P$, $P'$ and $Q$, $Q'$.

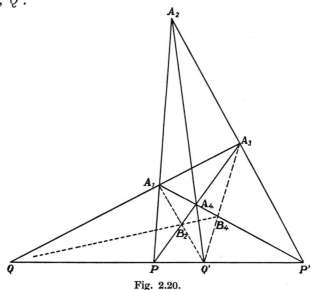

Fig. 2.20.

**Projection.**

**Definition.** The *projection* of a point $A$ from the centre $S$ on the line $m$, where $S \neq A$ and $S \notin m$, is the point $A' = SA \cap m$. Let $l$ and $m$ be two lines and $S$ a point outside $l$ and $m$. There is a one-to-one correspondence between the points of $l$ and their projections from $S$ on $m$. This correspondence is called a *perspectivity* with centre $S$.

REMARK. If we speak of the perspectivity between $l$ and $m$ with centre $S$, or of the projection of $l$ onto $m$ from $S$, it is always understood that $S \notin l$ and $S \notin m$.

**Theorem 2.4.5.** In $\mathfrak{P}(D_9)$: By a perspectivity harmonic pairs are transformed into harmonic pairs.

PROOF. Let $Q$, $Q'$ be harmonic with $P$, $P'$ on $l$; suppose first that $Q \neq Q'$. Let $X$, $X'$, $Y$, $Y'$ be the projections of $P$, $P'$, $Q$, $Q'$ from $S$ on $m$. The perspectivity from $l$ onto $m$ with centre $S$ is the product of two perspectivities with centre $S$, the first from $l$ onto $QY'$, the second from $QY'$ onto $m$. It suffices to prove that harmonic pairs are invariant under each of these perspectivities. Thus let $U$, $U'$, $Q$, $Y'$ be the projections of $P$, $P'$, $Q$, $Q'$ from $S$ on $QY'$. Let $T$ be $PU' \cap P'U$. The quadrangle $UU'ST$ shows that $T \in SQ'$. Then the quadrangle $PP'ST$ shows that $Q$, $Y'$ are harmonic with $U$, $U'$.

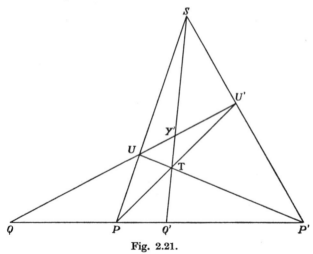

Fig. 2.21.

The proof for the perspectivity from $QY'$ onto $m$ is similar.

If $Q = Q'$, we consider first the perspectivity from $l$ onto $PX'$, then that from $PX'$ onto $m$. Let $P$, $X'$, $V$ be the projections of $P$, $P'$, $Q$ from $S$ on $PX'$. Let $Z = SP \cap P'V$. The quadrangle $SVX'Z$ shows that $ZX'$ passes through $Q'(=Q)$. Then the quadrangle $SP'QZ$ shows that $V$ is its own harmonic conjugate with respect to $P$, $X'$.

The second part of the proof is similar.

Theorems 2.4.2—2.4.5 show that the theory of harmonic pairs can be based entirely upon $D_9$; $D_{11}$, or even $D_{10}$, is not needed.

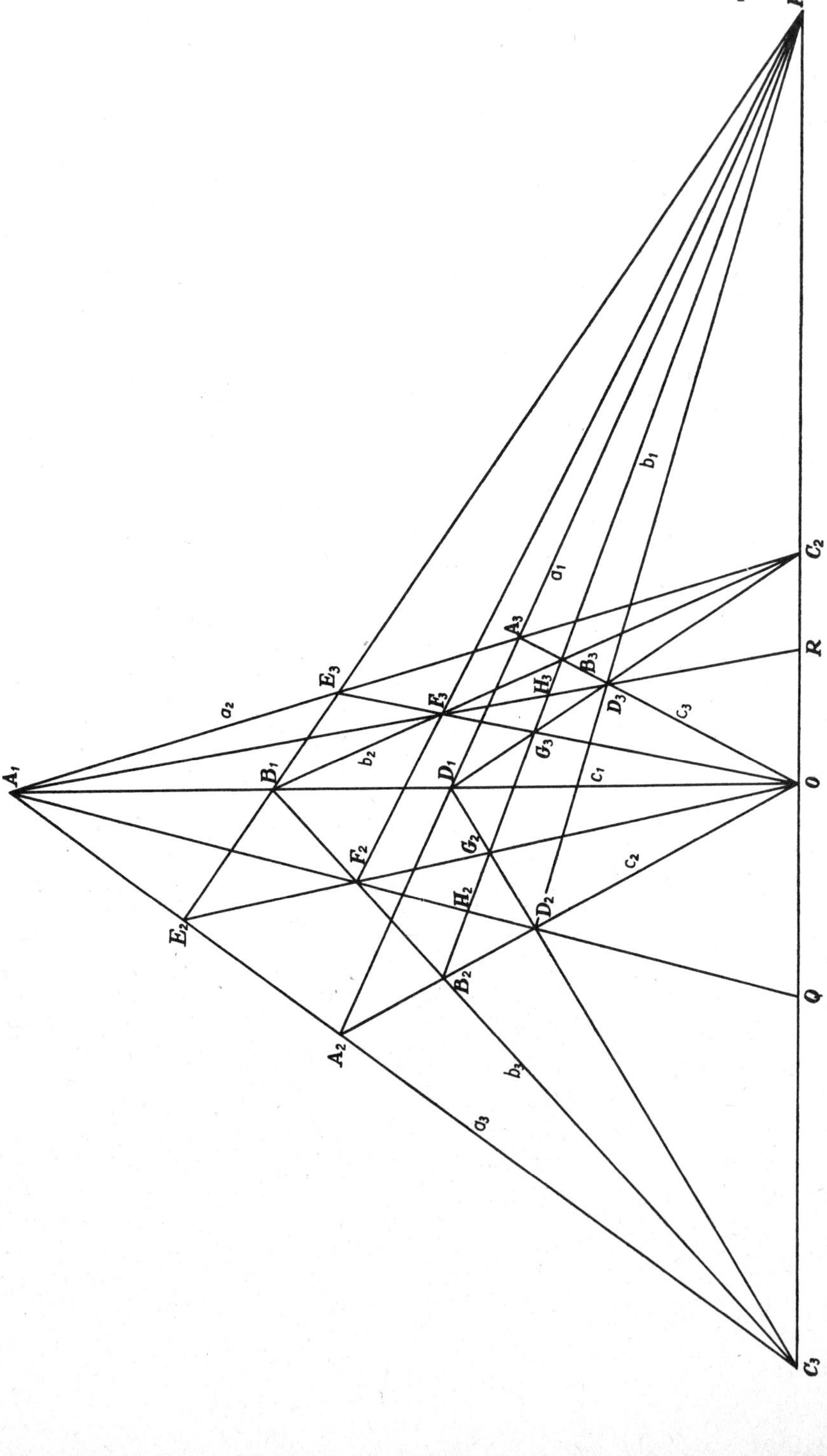
P
C2
R
O
Q
C3
A1
B1
D1
E3
F3
A3
H3
B3
D3
G3
F2
G2
H2
D2
E2
B2
A2
a1
a2
a3
b1
b2
b3
c1
c2
c3

However, the next theorem shows that $D_9$ is not much weaker than $D_{10}$. See the remark after Th. 2.5.6.

**Theorem 2.4.6.** In $\mathfrak{P}$: If the harmonic conjugate of $Q$ with respect to $P$, $P'$ is always different from $Q$, then $D_{10}$ is equivalent to $D_9$.

PROOF. It is more convenient to prove that $D_9$ implies $D_{10}^{I}$. In the figure of $D_{10}^{I}$ (in standard notation), denote $c_1 \cap a_1$ by $D_1$, $D_1 C_2 \cap c_3$ by $D_3$, $D_1 C_3 \cap c_2$ by $D_2$. $D_9^{II}$, applied to the triangles $A_1 A_2 A_3$ and $D_1 D_2 D_3$, shows that $P = D_2 D_3 \cap a_1$ lies on $l$. Now we see by the quadrangle $A_1 A_2 D_1 D_2$ that $Q = A_1 D_2 \cap l$ is the harmonic conjugate of $P$ with respect to $0$, $C_3$. Similarly, $R = A_1 D_3 \cap l$ is the harmonic conjugate of $P$ with respect to $0$, $C_2$. We introduce the additional points $E_3 = PB_1 \cap a_2$, $E_2 = PB_1 \cap a_3$, $F_2 = OE_2 \cap b_3$, $F_3 = OE_3 \cap b_2$. $D_9^{II}$, applied to the triangles $A_1 E_2 E_3$ and $B_1 F_2 F_3$, yields that $F_2 F_3$ contains $P$. Moreover, by the quadrangle $A_1 E_2 B_1 F_2$, $A_1 F_2$ contains $Q$; similarly, $A_1 F_3$ contains $R$, so that $A_1$, $F_2$, $D_2$ are collinear and $A_1$, $F_3$, $D_3$ are collinear. Finally we introduce the points $G_2 = D_1 D_2 \cap OE_2$, $G_3 = D_1 D_3 \cap OE_3$, $H_2 = B_2 G_2 \cap A_1 Q$, $H_3 = B_3 G_3 \cap A_1 R$.

The quadrangle $B_2 D_2 F_2 G_2$ shows that $B_2 G_2$ contains the harmonic conjugate of $Q$ with respect to $0$, $C_3$, that is $P$. Similarly, $B_3 G_3$ contains $P$. It remains to be shown that $B_2 G_2 = B_3 G_3$. To prove this, we remark that, by the quadrangle $B_2 G_2 O C_3$, $H_2$, $Q$ is harmonic with $D_2$, $F_2$; similarly $H_3$, $R$ is harmonic with $D_3$, $F_3$. Projecting the points $D_2 F_2 Q H_2$ from $P$ on $A_1 R$, we see by Th. 2.4.5 that $H_2 H_3$ contains $P$. But $H_2$ is on $B_2 G_2$ and $H_3$ is on $B_3 G_3$, which both contain $P$. It follows that $B_2 G_2 = B_3 G_3$, so that $B_2 B_3$ contains $P$.

Note the snake in the grass where we project the points of $A_1 Q$ from $P$ on $A_1 R$. This is only possible if $P$ is outside $A_1 Q$ and outside $A_1 R$, which we easily prove if $P \neq Q$ and $P \neq R$. Here the first hypothesis in the theorem is used. It is an unsolved problem whether $D_{10}$ is equivalent to $D_9$ in $\mathfrak{P}$ without any extra hypothesis.

## § 2.5. Projectivities between lines.

**Definition.** Let there be given lines $l_1, \ldots, l_{n+1}$ and points $S_1, \ldots, S_n$ such that $S_k \notin l_k$, $l_{k+1}$ $(k = 1, \ldots, n)$. We define a one-to-one mapping from $l_1$ onto $l_{n+1}$ as follows: let $X_1$ be any point of $l_1$; $X_{k+1} = S_k X_k \cap l_{k+1}$ $(k = 1, \ldots, n)$. A mapping which is constructed in this way is called a *projectivity* of order $n$, or a $\Pi_n$.

In particular a $\Pi_2$, denoted by $\varphi$, from $l$ onto itself is constructed as follows: choose $l' \neq l$ and $S_1$, $S_2$ outside $l$, $l'$. Then $\varphi A = (AS_1 \cap l')S_2 \cap l$. In other words, $\varphi$ results by projecting the points of $l$ from $S_1$ on $l'$ and by projecting back from $S_2$ on $l$. We call $S_1$ the first projection centre, $S_2$ the second projection centre and $l'$ the intermediate line. It is clear that $l \cap l' = O$ and $S_1 S_2 \cap l = U$ are invariant points of $\varphi$. We call $O$ the *first invariant point* and $U$ the *second invariant point* of $\varphi$.[1]

**Theorem 2.5.1.** In $\mathfrak{P}(D_{11})$, a projectivity $\varphi$ of order 2 from $l$ onto $l$ is uniquely determined by its first and second invariant points $O$, $U$ and the image $A'$ of one point $A$, provided $A$ and $A'$ are different from $O$ and $U$.

PROOF. A $\Pi_2$, say $\varphi$, with $\varphi A = A'$ and $O$, $U$ as first and second invariant points, is constructed as follows. Choose $l_1 \neq l$, through $O$, and $S_1$ outside $l$, $l_1$. $S_1 A \cap l_1 = A_1$, $A_1 A' \cap S_1 U = S_2$. For any point $X$ on $l$, $l_1 \cap S_1 X = X_1$ and $X' = \varphi X = X_1 S_2 \cap l$.

We prove first that $\varphi X$ is independent of the choice of $l_1$, then that it is independent of the choice of $S_1$; we may assume that $X \neq O$, $U$, $A$.

*a*). Choose $l_2 \neq l$, through $O$, not through $S_1$. $S_1 A \cap l_2 = A_2$, $S_1 X \cap l_2 = X_2$, $A_2 A' \cap S_1 U = S_3$. $X'' = \psi X = X_2 S_3 \cap l$. Apply the Generalized Desargues' Theorem to the points $A' A_1 A_2 | X' X_1 X_2 | O$ and the lines $S_1 A$, $A' A_2$, $A' A_1$, $| S_1 X$, $X' X_2$, $X' X_1, | l$, $l_1$, $l_2$. There exist points $P$, $Q$, $R$ and a line $m$ such that $P \in S_1 A$, $S_1 X$, $m$; $Q \in A' A_2$, $X' X_2$, $m$; $R \in A' A_1$, $X' X_1$, $m$. It follows that $P = S_1$, $R = S_2$, $m = S_1 U$, $Q = S_3$. Thus $X'' = X'$.

---

[1] It does not follow from the axioms that $O$ and $U$ are the only invariant points of a $\Pi_2$ which is not the identity. This question will play an important part in § 6.1.

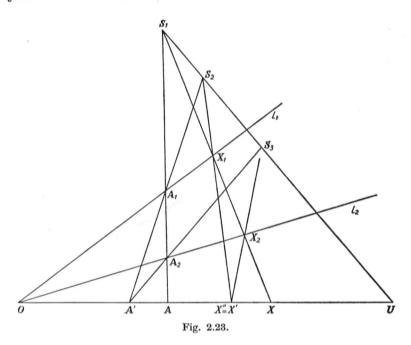

Fig. 2.23.

$b$). Choose $S_4$ outside $l$, $l_1$. $AS_4 \cap l_1 = A_4$, $A_4 A' \cap S_4 U = S_5$. $X''' = \chi X = (S_4 X \cap l_1) S_5 \cap l$. Let $\sigma$ be the central collineation $(S_1 S_2 U^2 l_1^2)$. $\sigma A = A'$, $\sigma S_4 = S_5$, $\sigma X = X'$. Thus $S_4 X$ and $S_5 X'$ intersect on $l_1$, and $X''' = X'$. (Fig. 2.24).

The $\Pi_2$ with $O$ as the first and $U$ as the second invariant point, and $A'$ as the image of $A$, will be denoted by $(A A' O^2 U^2)$.

**Exercise.** $O$ and $U$ need not be different. If they coincide, the reader can find a proof, analogous to that of Th. 2.5.1, of the following theorem:

**Theorem 2.5.2.** In $\mathfrak{P}(D_{10})$, a projectivity $\varphi$ of order 2 from $l$ onto $l$ with coinciding invariant points is uniquely determined by its invariant point $W$ and the image $A'$ of one point $A$, provided $A$, $A' \neq W$.

Instead of $(A A' W^2 W^2)$ we write $(A A' W^2)$.

**Theorem 2.5.3.** In $\mathfrak{P}(D_{11})$: The $\Pi_2$'s on $l$ with given first and second invariant points form a group.

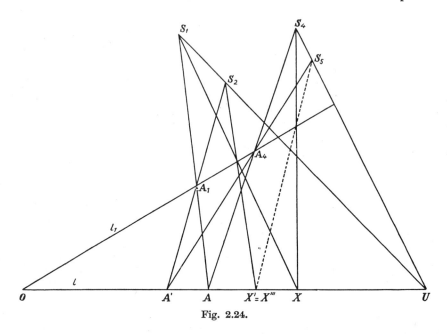

Fig. 2.24.

PROOF. Let $\varphi$ and $\psi$ be $\Pi_2$'s on $l$ with the same invariant points $O$, $U$. Choose for $\varphi$ a first projection centre $S_1$ and an intermediate line $l'$; construct the second projection centre $S_2$ as in the proof of Th. 2.5.1. Take $S_2$ as the first projection centre for $\psi$, $l'$ as the intermediate line and construct the second projection centre $S_3$. It is clear that the $\Pi_2$ with first projection centre $S_1$, second projection centre $S_3$ and intermediate line $l'$, is the transformation $\psi\varphi$. Thus the product of two $\Pi_2$'s with invariant points $O$, $U$ is again a $\Pi_2$, with the same invariant points. It is easily seen that the inverse of a $\Pi_2$ is a $\Pi_2$ too.

**Theorem 2.5.4.** In $\mathfrak{P}(D_{10})$: The $\Pi_2$'s on $l$ with given coinciding invariant points from a group.

PROOF as for the preceding theorem.

**Theorem 2.5.5.** In $\mathfrak{P}$: By every collineation a $\Pi_n$ between two lines is transformed into a $\Pi_n$ between their image lines.

PROOF. This is immediate from the definition of a $\Pi_n$ and the fact that a collineation transforms lines into lines and preserves the incidence relations.

REMARK. In order to apply Th. 2.5.5, we must know that collineations exist. $D_{11}$ is needed in order to secure the existence of a central collineation. If only special central collineations occur, $D_{10}$ suffices. This fact is used in the proof on Th. 2.5.7 below.

**Theorem 2.5.6.** In $\mathfrak{P}(D_{10})$: Every perspectivity between two lines can be extended to a special central collineation.

PROOF. Let the perspectivity $\pi$ between $l$ and $m$ have the centre $S$; we may suppose that $l \neq m$. Consider a point $A$ on $l$, not on $m$, and $\pi A = A'$. Let $d$ be the line joining $S$ to $l \cap m$. The special central collineation $(AA'S^2d^2)$ is an extension of $\pi$.

COROLLARY 2.5.6. In $\mathfrak{P}(D_{10})$: Every projectivity between two lines can be extended to a collineation.

PROOF. A projectivity $\rho$ is a product $\pi_k \ldots \pi_1$ of perspectivities $\pi_i$. Extend $\pi_i$ to a special central collineation $\mathfrak{C}_i$; then $\mathfrak{C}_k \ldots \mathfrak{C}_1$ is an extension of $\pi_k \ldots \pi_1$, which is obviously a collineation.

REMARK. By means of Th. 2.5.6, a simple proof of the following theorem can be given (by the same method as in the proof of Th. 2.5.5). In $\mathfrak{P}(D_{10})$: By a perspectivity harmonic pairs are transformed into harmonic pairs. It follows from Th. 2.4.6 that Th. 2.4.5 can be deduced from this under the extra hypothesis that the harmonic conjugate of $Q$ with respect to $P$, $P'$ is always different from $Q$.

**Theorem 2.5.7.** In $\mathfrak{P}(D_{10})$: By projection of $l$ onto $m$ from the centre $S$, every $\Pi_n$ of $l$ onto itself is transformed into a $\Pi_n$ of $m$ onto itself.

PROOF. Let $\varphi$ be the perspectivity from $l$ to $m$ with centre $S$. Extend $\varphi$ to a central collineation (Th. 2.5.6). Now Th. 2.5.7 follows immediately from Th. 2.5.5.

COROLLARY 2.5.7. In $\mathfrak{P}(D_{10})$: By a projectivity of $l$ onto $m$, every $\Pi_n$ of $l$ onto itself is transformed into a $\Pi_n$ of $m$ onto itself.

REMARK. It can be proved in $\mathfrak{P}(D_{11})$ that every $\Pi_n$ of $l$ onto $m$, where $l \neq m$, coincides with a $\Pi_2$. We do not need this theorem.

## § 2.6. Pappos' proposition.

The following statement is a theorem in $APPG$ (this will be shown after Th. 2.6.7; see $P'_{10}$) and is known as:

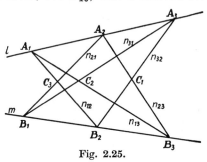

Fig. 2.25.

**Pappos' Proposition** $(P_{10})$. If $l$ and $m$ are different lines, $A_1$, $A_2$, $A_3$ different points on $l$, but not on $m$, $B_1$, $B_2$, $B_3$ different points on $m$, but not on $l$, then the points $C_1 = A_2 B_3 \cap A_3 B_2$ $C_2 = A_3 B_1 \cap A_1 B_3$, $C_3 = A_1 B_2 \cap A_2 B_1$ are collinear.

**Generalized Pappos' Proposition** $(P_{10}^*)$. If the 6 points $A_1 A_2 A_3 B_1 B_2 B_3$ and the 8 lines $n_{23} n_{31} n_{12} n_{32} n_{13} n_{21} \, l \, m$ are such that $A_i \in l$, $B_i \in m$ $(i = 1, 2, 3)$ and $A_i$, $B_k \in n_{ik}$ $(i, k = 1, 2, 3;$ $i \neq k)$, then there exist points $C_1$, $C_2$, $C_3$ and a line $n$ such that $C_1 \in n_{23}$, $n_{32}$, $n$; $C_2 \in n_{31}$, $n_{13}$, $n$; $C_3 \in n_{12}$, $n_{21}$, $n$. Note that there are no restrictions at all for the given points and lines: any number of them may coincide.

**Theorem 2.6.1.** In $\mathfrak{P}(P_{10})$, the generalized Pappos' proposition is a theorem.

PROOF. In every case where the conditions of $P_{10}$ are not fulfilled the assertion of the generalized Pappos' theorem is trivial.

REMARK. When applying $P_{10}^*$, we adopt a notational convention analogous to that in the case of $D_{11}^*$. Applying $P_{10}^*$ to the points $P_1, \ldots, P_6$ and the lines $q_1, \ldots, q_8$, means that $P_1$ plays the part of $A_1$ above, etc.

**Definition.** A *hexagon* is an ordered set of 6 different points $D_i$ and 6 lines $d_i$ $(i = 1, \ldots, 6)$ where $d_i = D_i D_{i-1}$ and $D_k \notin d_i$ if $k \neq i$, $i-1$ (subscripts mod 6). The lines $D_i D_{i+3}$ $(i = 1, 2, 3)$ are the (*main*) *diagonals* of the hexagon. The point $d_i \cap d_{i+3}$ is a *diagonal point* of the hexagon. A hexagon with collinear diagonal points is called a *Pascal hexagon*. The line on which its diagonal points lie is its *Pascal line*.

**Permutation Proposition.** If in a Pascal hexagon we interchange two adjacent vertices and if the resulting figure is a hexagon, then it is a Pascal hexagon.

**Weak Permutation Proposition.** If in a Pascal hexagon we interchange two vertices between which lies one other vertex and if the resulting figure is a hexagon, then it is a Pascal hexagon.

REMARKS. It follows from the Permutation Proposition that a Pascal hexagon changes into a Pascal hexagon by every permutation of the vertices (provided the result is a hexagon). From the Weak Permutation Proposition we can only infer that this is true for those permutations which leave the triples 1, 3, 5 and 2, 4, 6 invariant.

The name 'Pascal hexagon' alludes to Pascal's theorem in $APPG$; of course, this theorem can not be treated here because the notion of a conic has not been introduced.

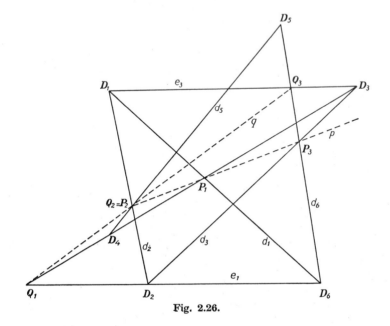

Fig. 2.26.

**Generalized Permutation Proposition.** Let there be given 9 points $D_1 D_2 D_3 D_4 D_5 D_6 P_1 P_2 P_3$ and 9 lines $d_1 d_2 d_3 d_4 d_5 d_6\ p\ e_1 e_3$, such that the conditions

   (i)    $D_i \in d_i,\ d_{i+1}$ ($i = 1, \ldots, 6$; subscripts mod 6);

   (ii)   $P_k \in d_k,\ d_{k+3},\ p$ ($k = 1, 2, 3$);

   (iii)  $D_2,\ D_6 \in e_1$; $D_1,\ D_3 \in e_3$

are fulfilled. Then there exist points $Q_1, Q_2, Q_3$ and a line $q$ such that $Q_1 \in e_1,\ d_4,\ q$; $Q_2 \in d_2,\ d_5,\ q$; $Q_3 \in e_3,\ d_6,\ q$.

**Theorem 2.6.2.** In $\mathfrak{P}(P_{10})$, the generalized permutation proposition is valid with the additional condition

   (iv) $p \neq d_2$ (the Pascal line is different from the line through the points interchanged).

PROOF. Apply $P_{10}^*$ to the points $D_1 P_1 D_6 P_3 D_2 D_3$ and the lines $d_4 d_6 d_2 e_1 e_3 p d_1 d_3$. We find points $C_1, C_2, C_3$ and a line $n$ such that $C_1 \in d_4,\ e_1,\ n$; $C_2 \in e_3,\ d_6,\ n$; $C_3 \in d_2,\ p,\ n$. Choose $Q_1 = C_1,\ Q_3 = C_2$, $q = n,\ Q_2 = C_3$. By (iv), $Q_2 = P_2$, so $Q_2 \in d_5$, which proves the theorem.

It is convenient to replace condition (iv) by a condition that does not refer to $p$. Therefore we shall prove:

**Theorem 2.6.3.** In Th. 2.6.2, condition (iv) can be replaced by (v) $D_2 \neq D_6$ and $D_1 \neq D_3$.

PROOF. We may suppose that $d_2 = p$.

First suppose that $D_3 \notin p$; then $d_3 \neq p$. $D_2$ as well as $P_3$ belongs to both $d_3$ and $p$, so $D_2 = P_3$; then $D_2 \in d_6$ and by (v) $d_6 = e_1$. In the same way, if $D_6 \notin p$, then $d_4 = e_3$.

Now we distinguish between three cases a), b), c).

**Case a).** $D_3 \notin p$, $D_6 \notin p$. Now $d_6 = e_1$ and $d_4 = e_3$, so we may take $Q_1 = Q_3$; then $Q_2$ and $q$ can always be found so as to satisfy the theorem.

**Case b).** $D_3 \notin p$, $D_6 \in p$. As $d_2 = p$, we have also $D_2 \in p$, and $e_1 = p = d_2$. Moreover $d_6 = e_1$, so we can find $Q_1, Q_2, Q_3$ on $p$ to satisfy the theorem.

**Case c).** $D_3,\ D_6 \in p$. Now $e_1 = p$, $e_3 = p$, $d_2 = p$. Find $Q_1, Q_2, Q_3$ on $p$ to satisfy the theorem.

**Generalized Weak Permutation Proposition.** Let there be given 9 points $C_1 C_2 C_3 C_4 C_5 C_6 P_1 P_2 P_3$ and 9 lines $c_1 c_2 c_3 c_4 c_5 c_6 p f_1 f_4$, such that the conditions

(i)    $C_i \in c_i$, $c_{i+1}$ ($i = 1, \ldots, 6$; subscripts mod 6);

(ii)    $P_k \in c_k$, $c_{k+3}$, $p$ ($k = 1, 2, 3$);

(iii)    $C_3$, $C_6 \in f_1$; $C_1$, $C_4 \in f_4$

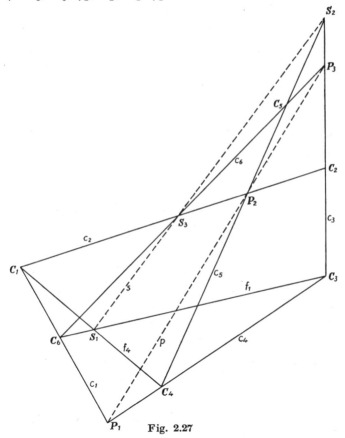

Fig. 2.27

are fulfilled. Then there exist points $S_1$, $S_2$, $S_3$ and a line $s$ such that $S_1 \in f_1$, $f_4$, $s$; $S_2 \in c_3$, $c_5$, $s$; $S_3 \in c_2$, $c_6$, $s$.

**Theorem 2.6.4.** In $\mathfrak{P}(P_{10})$, the Generalized Weak Permutation Proposition holds under the conditions $C_1 \neq C_4$, $C_3 \neq C_6$, $C_1 \neq C_2$, $C_3 \neq C_2$.

PROOF. We apply the Generalized Permutation theorem three times, first interchanging $C_1$ with $C_2$, then $C_1$ with $C_3$, finally $C_2$ with $C_3$. In other words, the points and lines $D_1D_2D_3D_4D_5D_6P_1P_2P_3d_1d_2d_3d_4d_5d_6pe_1e_3$ in the formulation of the generalized permutation proposition above (fig. 2.26) are replaced successively by:

(step 1) $C_1\,C_2\,C_3\,C_4\,C_5\,C_6\,P_1\,P_2\,P_3\,c_1\,c_2\,c_3\,c_4\,c_5\,c_6\,p\,e_1\,e_3$,
  giving $Q_1,\,Q_2,\,Q_3,\,q$;
(step 2) $C_1C_3C_4C_5C_6C_2Q_2Q_3Q_1c_2e_3c_4c_5c_6e_1qc_3f_4$,
  giving $K_1,\,K_2,\,K_3,\,r$;
(step 3) $C_2\,C_3\,C_1\,C_4\,C_5\,C_6\,R_3\,R_1\,R_2\,e_1\,c_3\,e_3\,f_4\,c_5\,c_6\,r\,f_1\,c_2$,
  giving $S_1,\,S_2,\,S_3,\,s$,

which satisfy the theorem.

According to Th. 2.6.3, we had to impose the following conditions: for step 1, $C_1 \ne C_3$, $C_2 \ne C_6$; for step 2, $C_1 \ne C_4$, $C_3 \ne C_2$; for step 3, $C_2 \ne C_1$, $C_3 \ne C_6$. However, the conditions $C_1 \ne C_3$ and $C_2 \ne C_6$ are unnecessary.

In order to prove this, we have to consider two cases a) and b).

**Case a).** $C_1 = C_3$; $C_1 \ne C_2$, $C_4$, $C_6$. Now $c_2 = c_3$, $f_4 = c_4$, $c_1 = f_1$, so we can take $S_1 = P_1$, $S_2 = P_2$, $S_3 = P_3$.

**Case b).** $C_2 = C_6$; $C_2 \ne C_1$, $C_3$. Now $c_2 = c_1$, $c_3 = f_1$. $C_2 \in c_1$, $c_2$, $c_3$, $c_6$, so we may take $S_3 = C_2$; as $S_1$ and $S_2$ are on $f_1 = c_3$, we can take $s = c_3$.

**Theorem 2.6.5. (Hessenberg's Theorem).** In $\mathfrak{P}(P_{10})$, Desargues' Proposition $D_{11}$ is valid.

PROOF. Let $A_1A_2A_3$ and $B_1B_2B_3$ be triangles satisfying the conditions of $D_{11}$. $A_1B_1$, $A_2B_2$, $A_3B_3$ pass through $O$, $a_i \cap b_i = C_i$.

Apply the Generalized Weak Permutation theorem to the points $B_1D_2A_2B_2D_5A_1OB_3A_3$ and the lines $c_1b_2a_1c_2b_1a_2c_3a_3b_3$, where $D_2 = a_1 \cap b_2$, $D_5 = a_2 \cap b_1$ (the case where $a_1 = b_2$ or $a_2 = b_1$ is trivial). We find points $S_1$, $S_2$, $S_3$, and a line $s$, such that $S_1 \in a_3$, $b_3$, $s$; $S_2 \in a_1$, $b_1$, $s$ and $S_3 \in b_2$, $a_2$, $s$. This proves $D_{11}$.

The Generalized Weak Permutation Theorem can be applied if $B_1 \ne B_2$, $A_2 \ne A_1$, $B_1 \ne D_2$, $A_2 \ne D_2$. The first two conditions are satisfied for triangles. If we assume $B_1 \notin a_1$, $A_2 \notin b_2$, then the

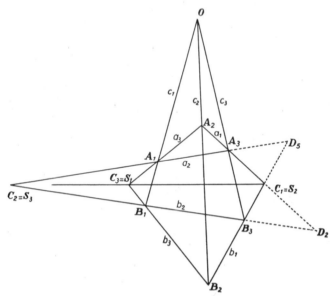

Fig. 2.28.

other two conditions are fulfilled. We may suppose $a_1 \neq b_2$ and $a_2 \neq b_1$. The cases $B_1 \,\epsilon\, a_1$ and $A_2 \,\epsilon\, b_2$ remain to be examined.

Of course, it is sufficient that for some subscripts $i$, $k$ $(i \neq k)$ $B_i \notin a_i$, $A_k \notin b_k$. There are two cases in which such subscripts $i$, $k$ cannot be found, namely

I) $A_i \,\epsilon\, b_i$ $(i = 1, 2, 3)$ or $B_i \,\epsilon\, a_i$ $(i = 1, 2, 3)$;

II) $A_1 \,\epsilon\, b_1$, $A_2 \,\epsilon\, b_2$, $B_1 \,\epsilon\, a_1$, $B_2 \,\epsilon\, a_2$ (or the same after permuting the subscripts 1, 2, 3).

In case I) $OA_1A_2$ and $B_3C_2C_1$ are triangles which are not in case I) or II); application of $D_{11}$ to these triangles proves the theorem. Case II) implies either $A_3 = B_3$ or $a_3 = b_3$, which contradicts the hypothesis of the theorem. (See Appendix 1.)

This completes the proof of Th. 2.6.5.

**Second Quadrangle Proposition** $(Q_2)$. Let $A_1A_2A_3A_4$ and $B_1B_2B_3B_4$ be two complete quadrangles and $l$ a line which contains none of the points $A_i$, $B_i$. If $A_1A_2 \cap l = B_3B_4 \cap l$, $A_1A_3 \cap l = B_2B_4 \cap l$, $A_1A_4 \cap l = B_2B_3 \cap l$, $A_2A_3 \cap l = B_1B_4\cap l$, $A_2A_4 \cap l = B_1B_3 \cap l$, then $A_3A_4 \cap l = B_1B_2 \cap l$.

REMARK. If we call $A_1A_2$ and $B_3B_4$ anticorresponding sides, then the theorem says: If five pairs of anticorresponding sides intersect $l$ in the same point, then the same is true for the sixth pair.

**Theorem 2.6.6.** $Q_2$ holds in $\mathfrak{P}(P_{10})$.

PROOF. First of all, we treat the case where $A_1 = B_3$ and $A_2 = B_4$. Let $a_{ik} = A_iA_k$, $b_{ik} = B_iB_k$, $P_{ik} = a_{ik} \cap l$. Now $B_1 = A_1P_{24} \cap A_2A_3$; $B_2 = A_2P_{13} \cap A_1A_4$.

If $P_{24} = P_{13}$, then $B_1 = A_3$ and $B_2 = A_4$, so the theorem is immediate.

If $P_{24} \neq P_{13}$, we apply $P_{10}^*$ to the points $A_2A_3B_1A_1B_2A_4$

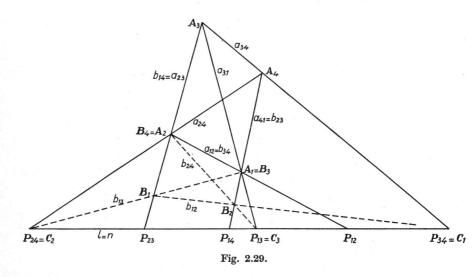

Fig. 2.29.

and the lines $a_{34}b_{13}b_{24}b_{12}a_{24}a_{13}a_{23}a_{14}$. We find points $C_1$, $C_2$, $C_3$ and a line $n$ such that $C_1 \in a_{34}$, $b_{12}$, $n$; $C_2 \in b_{13}$, $a_{24}$, $n$; $C_3 \in b_{24}$, $a_{13}$, $n$. Thus $C_2 = P_{24}$, $C_3 = P_{13}$, $n = l$, $C_1 = B_1B_2 \cap l = A_3A_4 \cap l = P_{34}$.

Now we consider the general case. If we put $D_1 = A_1P_{24} \cap A_2A_3$, $D_2 = A_2P_{13} \cap A_1A_4$, $D_3 = A_1$, $D_4 = A_2$, then $D_1D_2D_3D_4$ is a quadrangle satisfying the conditions of the special case above; it follows that $D_hD_i \cap l = A_jA_k \cap l$ for every permutation $h$, $i$, $j$, $k$ of the subscripts 1, 2, 3, 4. Because of Th. 2.6.5 and

Th. 2.4.1, we may now apply $Q_1$ to the quadrangles $D_1D_2D_3D_4$ and $B_1B_2B_3B_4$, which gives the desired result.

**Theorem 2.6.7.** (Converse of Th. 2.6.6.) $P_{10}$ holds in $\mathfrak{P}(Q_2)$.

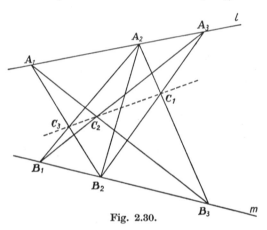

Fig. 2.30.

PROOF. In order to prove $P_{10}$ for the points $A_1A_2A_3B_1B_2B_3$ it suffices to apply $Q_2$ to the quadrangles $A_1A_2B_2B_3$ and $B_2B_1A_3A_2$.

**Special cases of Pappos' Proposition.**

In the figure of $P_{10}$ we consider the hexagon $A_1B_3C_1A_3B_1C_3$. Its diagonals (connecting opposite vertices) are $A_1A_3$, $B_1B_3$, $C_1C_3$; its diagonal points (intersections of opposite sides) are $A_2$, $B_2$, $C_2$. The diagonal point $A_2$ is said to correspond to the diagonal $A_1A_3$, etc. Thus $P_{10}$ can be formulated as follows:

($P'_{10}$). If in a hexagon two diagonal points are on the corresponding diagonals, the third diagonal point lies on its corresponding diagonal too. (In this form Pappos' theorem was proved in *APPG*; Th. 1.4.6).

This proposition is self-dual. Therefore we have:

**Theorem 2.6.8.** The duality principle is valid in $\mathfrak{P}(P_{10})$.

The following special cases are interesting:

$P_9 = P'_{10}$ with the additional condition that the three diagonals pass through one point.

$dP_9 = P'_{10}$ with the additional condition that the three diagonal points are on one line. ($dP_9$ is the dual of $P_9$).

**Theorem 2.6.9.** $P_9$ and $dP_9$ are valid in $\mathfrak{P}(D_9)$.

PROOF. We prove $P_9$; $dP_9$ follows by duality.

Let the hexagon $A_1B_3C_1A_3B_1C_3$ satisfy the hypothesis of $P_9$.

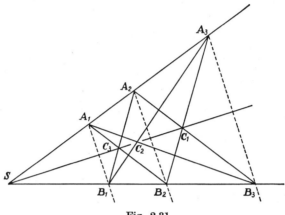

Fig. 2.31.

Let $S$ be the common point of the diagonals $A_1A_3$, $B_1B_3$, $C_1C_3$, By applying $dD_9^I$ to the triangles $B_2A_1A_3$ and $A_2B_1B_3$, we find that $A_1B_1$, $A_2B_2$, $A_3B_3$ are concurrent. Then $D_9^I$, applied to the triangles $B_3A_1A_2$ and $A_3B_1B_2$, discloses that $C_2 \in SC_1$; thus $C_1$, $C_2$, $C_3$ are collinear.

Theorem 2.6.9 strongly contrasts with the fact that $P_{10}$ is independent of $V_1$, $V_2$, $V_3$, $D_{11}$; this will be proved in Th. 3.5.2.

Let us now return to the general proposition $P_{10}$. Its importance in the axiomatics of projective geometry is expressed by the next theorem. I recall (Th. 2.5.3) that in $\mathfrak{P}(D_{11})$, the projectivities of order 2 on $l$ with given first and second invariant points form a group; by Th. 2.6.5, this is also true in $\mathfrak{P}(P_{10})$.

**Theorem 2.6.10.** In $\mathfrak{P}(P_{10})$, the group of $\Pi_2$'s on $l$ with given invariant points $O$, $U$ is abelian.

PROOF. Let $\varphi$, $\psi$ be two $\Pi_2$'s on $l$ with the same invariant points $O$, $U$. We must prove that $\varphi\psi = \psi\varphi$. This time we choose the same centres $S_1$ and $S_2$ for $\varphi$ and $\psi$; let $l'$ and $l''$ be the intermediate lines for $\varphi$ and $\psi$ respectively. Let $A(\neq O, U)$ be any point on $l$. $S_1A \cap l' = B$, $S_1A \cap l'' = G$, $S_2B \cap l = C = \varphi A$, $S_2G \cap l = H = \psi A$,

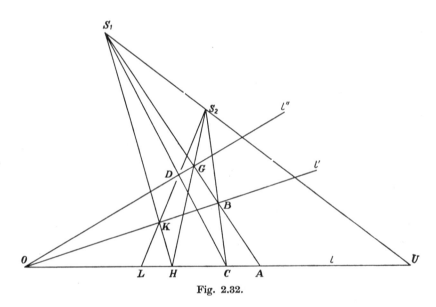

Fig. 2.32.

$S_1 H \cap l' = K$, $S_2 K \cap l = L = \varphi\psi A$, $S_1 C \cap l'' = D$, $S_2 D \cap l =$ $= F = \psi\varphi A$. We must prove that $L = F$, or, what amounts to the same, that $S_2$, $K$, $D$ are collinear. In the hexagon $BCDGHK$ the diagonal points $S_1$, $O$ are on the corresponding diagonals $BG$, $CH$; thus by $P'_{10}$, $S_2$ is on $DK$.

**Theorem 2.6.11.** In $\mathfrak{P}(D_{10})$, the group of $\Pi_2$'s on $l$ for which the invariant points coincide in $O$, is abelian.

PROOF. In the preceding proof the diagonal points $S_1$, $S_2$, $O$ are now collinear, so that we can use $dP_9$ instead of $P'_{10}$; by Th. 2.6.9, $dP_9$ is valid in $\mathfrak{P}(D_9)$. Th. 2.5.4 is used.

REMARK. If $\rightarrow$ means "implies in $\mathfrak{P}$" and $\leftrightarrow$ means "is equivalent in $\mathfrak{P}$ to", we can summarize some results of this chapter like this:

$$P_{10} \rightarrow D_{11} \rightarrow D_{10} \rightarrow D_9 \rightarrow P_9;\ P_{10} \leftrightarrow Q_2;\ D_{11} \leftrightarrow Q_1.$$

It is not known whether $P_9 \rightarrow D_9$; we shall see later that $D_{10} \nleftrightarrow D_{11} \nleftrightarrow P_{10}$, where $\nleftrightarrow$ means: "does not imply in $\mathfrak{P}$" (Th. 3.5.1, 3.5.2). It is not known whether $D_9 \rightarrow D_{10}$ is generally true, but it can be proved as a consequence of the harmonic proposition (Th. 2.4.6).

# CHAPTER III

# COORDINATES IN THE PLANE

## § 3.1. Ternary fields attached to a given projective plane.

In the projective plane $\mathfrak{P}_0$ (V1, V2, V3) we choose four points $O$, $X$, $Y$, $E$, of which no three are collinear. If $P$ is a point not on $XY$, the points $P_1 = OE \cap YP$ and $P_2 = OE \cap XP$ are

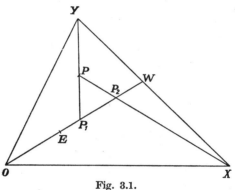

Fig. 3.1.

uniquely determined. Conversely, if $P_1$ and $P_2$ are given on $OE$, but not on $XY$, then $P = YP_1 \cap XP_2$ is uniquely determined. We have thus a one-to-one correspondence between $\mathfrak{P}_0 \setminus XY$ and $(OE \setminus \{W\}) \times (OE \setminus \{W\})$, where $W = OE \cap XY$. $\mathfrak{P}_0 \setminus XY$ will be called the *affine plane*.

**Definition.** $P_1$ and $P_2$ are the *coordinates* of $P$ in the *coordinate system $OXYE$*.

We shall denote coordinates (points on $OW$ but $\neq W$) by lower case letters; $x$ and $y$ are variables for the first and second coordinate of a point, respectively. Instead of $O$ and $E$, considered as coordinates, we write 0 and 1. The point with coordinates $a$ and $b$ is denoted by $(a, b)$.

**Exercise.** The reader should verify the following statements from the definitions: $O = (0, 0)$; $E = (1, 1)$; $x = c$ is the equation of a line through $Y$; $x = 0$ that of $OY$; $y = c$ is the equation of a line through $X$; $y = 0$ that of $OX$; $x = y$ is the equation of $OE$.

Let a line $l$ be given, which does not contain $Y$. We define $m$ and $n$ by

(1) $$l \cap XY = C;$$
(2) $$l \cap OY = B = (0, n);$$
(3) $$(CO \cap EY)X \cap OE = m; \text{ and thus}$$
(4) $$CO \cap EY = (1, m).$$

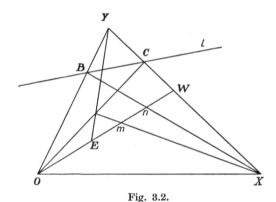

Fig. 3.2.

$l$ determines $m$ and $n$ uniquely. Conversely, if $m$ and $n$ are given, we construct $l$ as follows:

(5) $$B = (0, n); \quad (Xm \cap YE)O \cap XY = C; \quad BC = l.$$

We call $m$, $n$ the coordinates of $l$, and write $l = [m, n]$.

Examples: I. If $m = 0$, then $X \in l$; equation $y = n$ ($XB$ in fig. 3.2).

II. If $n = 0$, then $O \in l$ ($OC$ in fig. 3.2).

III. If $m = 1$, $n = 0$, then $l = OE$; equation $y = x$.

REMARK. In fig. 3.2, $C$ depends only upon $m$; the lines through a point of $XY$ have the same coordinate $m$. If $l = [m, n]$, we shall denote $l \cap XY$ by $(m)$.

Let $(x, y)$ be any point on $l$. Given $x$, we construct $y$ by

(6) $$y = (Yx \cap l)X \cap OE. \qquad \text{(Fig. 3.3).}$$

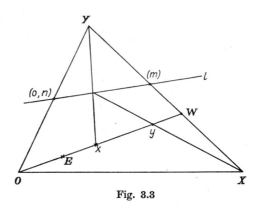

Fig. 3.3

Thus $y$ is a function of $x$, depending also upon $m$, $n$:

(7)                           $y = \Phi(x, m, n).$

(7) can be considered as the equation of the line $l = [m, n]$.

We begin by deriving some simple properties of the function $\Phi$, which are consequences of V1, V2, V3. As we saw a moment ago, $[0, n]$ has the equation $y = n$; this fact is expressed by

(i)                           $\Phi(a, 0, n) = n.$

$[1, 0]$ has the equation $y = x$, that is

(ii)                          $\Phi(a, 1, 0) = a.$

If $(0, y)$ is a point on $[m, n]$, we have by (2), $y = n$. Thus

(iii)                         $\Phi(0, m, n) = n.$

Let $P = (1, y)$ be a point on $\lceil m, 0\rceil$. By (5) and (3) we have

$$y = (YE \cap l)X \cap OE = (YE \cap CO)X \cap OE = m.$$

(iv)                          $\Phi(1, m, 0) = m.$

Let $l$ be the line $[m, n]$. If $m \neq 0$, then $X \notin l$, so $l$ has a unique point of intersection with $y = b$. It follows that

(v)   if $m \neq 0$, then $\Phi(x, m, n) = b$ has a unique solution for $x$.

Let a point $P = (a, b)$ and a point $B = (0, n)$ be given. If $a \neq 0$, then $P \notin OY$, so the line $PB = l$ is unique and does not contain $Y$. Thus there is a unique value for $m$ such that $l = [m, n]$. It follows that

(vi)   if $a \neq 0$, then $\Phi(a, z, n) = b$ has a unique solution for $z$.

Let a point $P = (a, b)$ and a point $m$ on $OE$ be given. We construct $C$ as in (5), and $l = PC$. $l = [m, n]$ for just one value of $n$, and $\Phi(a, m, n) = b$. Thus

(vii)        $\Phi(a, m, u) = b$ has a unique solution for $u$.

Let $A = (a, b)$ and $A' = (a', b')$ be any points such that $a \neq a'$. Then $Y \notin AA'$, so there are uniquely determined values for $m$ and $n$ such that $AA' = [m, n]$. Thus

(viii)   the equations $\Phi(a, z, u) = b$ and $\Phi(a', z, u) = b'$, where $a \neq a'$, have a unique solution for $z$ and $u$.

Let $l = [m, n]$ and $l' = [m', n']$ be any lines such that $m \neq m'$; then the point $S = l \cap l'$ is not on $XY$; the coordinates of $S$ satisfy the equations of $l$ and $l'$. This gives

(ix)   if $m \neq m'$, the equation $\Phi(x, m, n) = \Phi(x, m', n')$ has a unique solution for $x$.

These properties are not independent. We derive (vi) by taking $a' = 0$, $b' = n$ in (viii) and then applying (iii). Moreover, if we substitute $m' = 0$, $n' = b$ in (ix) and then apply (i), we obtain (v).

**Definition.** A *ternary field* is a class which contains at least two different elements 0 and 1, and in which a function $\Phi(x, y, z)$ is defined satisfying (i), (ii), (iii), (iv), (vii), (viii), (ix).

As we remarked just now, (v) and (vi) hold in every ternary field. (See Appendices 2 and 3).

**Theorem 3.1.1.** If in a projective plane $\mathfrak{P}_0$ a coordinate system $OXYE$ is given, the points of $OE$ different from $W = OE \cap XY$ form a ternary field, in which the function $\Phi$ is defined by

(8) $\Phi(x, m, n) = (\{(Xn \cap OY)[(Xm \cap YE)O \cap XY]\} \cap Yx)X \cap OE$.

PROOF. This is left to the reader, as most of it has been shown above. See fig. 3.4.

The ternary field mentioned in Th. 3.1.1 will be said to be *attached* to the projective plane $\mathfrak{P}_0$ by the coordinate system $OXYE$. Any ternary field attached to a projective plane $\mathfrak{P}_0((\mathfrak{A})$, where $\mathfrak{A}$ is some set of axioms, will be denoted by $\mathfrak{T}(\mathfrak{A})$. The axioms V1, V2, V3 will always be included in $\mathfrak{A}$ even if they are not mentioned; $\mathfrak{T}$ (V1, V2, V3) will be abbreviated to $\mathfrak{T}$. Thus, e.g., Th. 3.2.1, as formulated below, is an abbreviated form of the

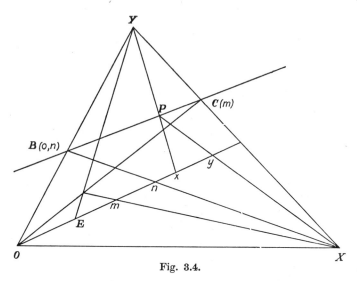

Fig. 3.4.

theorem: In any ternary field attached to a projective plane in which $D_{10}$ is valid, the elements form an abelian group under addition. Analogously for theorems on $\mathfrak{T}(A_1, \ldots, A_k)$, where $A_1, \ldots, A_k$ are axioms.

## § 3.2. Ternary field and axiom system.[1])

We are going to study the connection between $\mathfrak{T}(\mathfrak{A})$ and the underlying axiom system $\mathfrak{A}$. First of all, we define an addition in an arbitrary ternary field.

**Definition.** The *sum* $x+z$ of two elements $x$, $z$ of a ternary field is $\Phi(x, 1, z)$.

In view of (ii) and (iii) we have $x + 0 = x$; $0 + z = z$.

In $\mathfrak{T}$ (V1, V2, V3) we have as a special case of (8) the following construction for $x + z$:

(9)   $x + z = [(Xz \cap OY)W \cap Yx]X \cap OE$ (fig. 3.5).

For a given $z$, this is the construction of a $\varPi_2$ with intermediate

---

[1]) In the figures of this section the coordinate system $OXYE$ is represented by a Cartesian system, in which $X$, $Y$ are the points at infinity of the axes; $W$ is then the point at infinity of $OE$. As to the significance of these figures we remind the reader of the remark at the end of section 1.1.2.

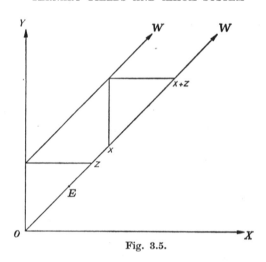

Fig. 3.5.

line $(Xz \cap OY)W$ and projection centres $Y$, $X$. Its first and second invariant point coincide in $W$.

We now suppose that $D_{10}$ is valid. Then, by Th. 2.5.2, this $\Pi_2$ is defined by its invariant point $W$ and the image $z$ of $O$; it is denoted by $(OzW^2)$. Thus, $x + z = y$ if and only if $(OzW^2)x = y$, or, what comes to the same, $(OzW^2)=(xyW^2)$. Because in $\mathfrak{P}(D_{10})$ the $\Pi_2$'s with given unique invariant point $W$ form an abelian group (Th.2.6.11), $(OzW^2)=(xyW^2)$ is equivalent to $(OzW^2)(OxW^2)$ $=(OyW^2)$. Thus the mapping $x \to (OxW^2)$ is an isomorphism between the points of $OE$ ($W$ excepted) and the $\Pi_2$'s on $OE$ with $W$ as the only invariant point. This proves

**Theorem 3.2.1.** In $\mathfrak{T}(D_{10})$, the elements form an abelian group under addition.

**Definition.** The *product* of two elements $x$, $z$ of a ternary field is $x \cdot z = \Phi(x, z, 0)$. For $x \cdot z$ we also write $xz$.

In view of (ii) and (iv) we have $x \cdot 1 = 1 \cdot x = x$; in view of (i) and (iii) $x \cdot 0 = 0 \cdot x = 0$.

The following construction for $x \cdot z$ in $\mathfrak{T}(V1, V2, V3)$ is a special case of (8):

(10)     $x \cdot z = [(Xz \cap YE)O \cap Yx]X \cap OE$ (fig. 3.6).

For a given $z \neq 0$ this is the construction of a $\Pi_2$ with intermediate line $(Xz \cap YE)O$ and projection centres $Y$, $X$.

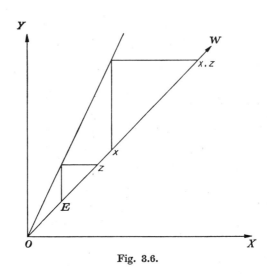

Fig. 3.6.

If $D_{11}$ is valid, then by Th. 2.5.1 this $\Pi_2$ is uniquely determined by its invariant points $O$, $W$ and the image $z$ of $E$; it is denoted by $(EzO^2W^2)$. Thus, $x \cdot z = t$ if and only if $(EzO^2W^2)x = t$, or, what amounts to the same, $(EzO^2W^2)=(xtO^2W^2)$. Moreover, in $\mathfrak{P}(D_{11})$ the $\Pi_2$'s with given invariant points $O$, $W$ form a group, so $(EzO^2W^2)=(xtO^2W^2)$ is equivalent to $(EzO^2W^2)(ExO^2W^2)=(EtO^2W^2)$. Thus the mapping $x \to (ExO^2W^2)$ is an anti-isomorphism between the points of $OE$ ($O$ and $W$ excepted) and the $\Pi_2$'s on $OE$ with $O$ and $W$ as invariant points. This proves

**Theorem 3.2.2.** In $\mathfrak{X}(D_{11})$, the elements $\neq 0$ form a group under multiplication.

In view of Th. 2.6.10, we have

**Theorem 3.2.3.** In $\mathfrak{X}(P_{10})$, the elements $\neq 0$ form an abelian group under multiplication.

**Theorem 3.2.4.** In $\mathfrak{X}(D_{10})$, we have $\Phi(x, m, n) = x \cdot m+n$.

PROOF. We follow the construction of $x \cdot m + n$.

$$p = x \cdot m = [(Xm \cap YE)O \cap Yx]X \cap OE.$$

In fig. 3.7, $Xm \cap YE = A$, $AO \cap Yx = R$, $RX \cap OE = p$.

$$q = p + n = [(Xn \cap OY)W \cap Yp]X \cap OE.$$
$$Xn \cap OY = B, \ BW \cap Yp = Q, \ QX \cap OE = q.$$

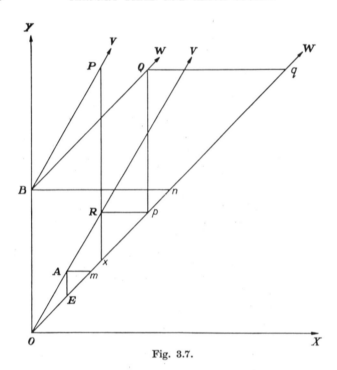

Fig. 3.7.

On the other hand, the construction of $\Phi(x, m, n)$ runs as follows:

$$\Phi(x, m, n) = (\{(Xn \cap OY)[(Xm \cap YE)O \cap XY]\} \cap Yx)X \cap OE =$$
$$= \{B(AO \cap XY) \cap Yx\}X \cap OE.$$

$$AO \cap XY = V, \quad BV \cap Yx = P.$$

It must be proved that $PX \cap OE = q$, or, what comes to the same, that $PX = QX$. This is a consequence of the existence of the special central collineation $\mathfrak{C} = (OBY^2\overline{XY}^2)$. We have $\mathfrak{C}(OV) = BV$; $\mathfrak{C}R = P$; $\mathfrak{C}(RX) = PX$; $\mathfrak{C}(OW) = BW$; $\mathfrak{C}p = Q$; thus $Q \epsilon PX$.

The following theorem will be useful for the proofs of further properties of the function $\Phi$.

**Theorem 3.2.5.** In $\mathfrak{P}(D_{10})$: If $\varphi$ is a projectivity of $OW$ onto itself which leaves $O$ and $W$ invariant, then for every $x$ and $y$ $\varphi(x + y) = \varphi(x) + \varphi(y)$ holds.

PROOF. As we saw before, it follows from (9) that $x + y = (OyW^2)x$. Denote $\varphi(x)$ bij $x'$. Applying Th. 2.5.7 we find that $\varphi$ transforms $(OyW^2)$ into $(Oy'W^2)$. Thus $\varphi(x + y)=(Oy'W^2)x' = x' + y' = \varphi(x) + \varphi(y)$.

**Theorem 3.2.6.** In $\mathfrak{X}(D_{10})$, we have $(x+p)m = xm+pm$ (distributively from the right).

PROOF. As a consequence of (10), for fixed $m \neq 0$ the transformation $\psi$ defined bij $\psi(x) = xm$ is a projectivity.

Thus, by Th. 3.2.5, $\psi(x + p) = \psi(x) + \psi(p)$, that is $(x + p)m = xm + pm$.[1])

**Theorem 3.2.7.** In $\mathfrak{X}(D_{10})$, $x(m+q) = xm+xq$.

PROOF. It follows from (10) that the transformation $\varphi$ which, for fixed $x \neq 0$, transforms $m$ into $\varphi(m) = xm$, is a $\Pi_3$. In fact, we obtain $xm$ from $m$ by projecting first from $X$ onto $YE$, then from $O$ onto $Yx$, and finally from $X$ onto $OE$ (fig. 3.6).

Then, by Th. 3.2.5, $\varphi(m + q) = \varphi(m) + \varphi(q)$, i.e. $x(m + q)= xm + xq$.

**Theorem 3.2.8. (Summary).**

$\mathfrak{X}$ (V1, V2, V3) is a ternary field.

$\mathfrak{X}(D_{10})$ is an abelian group under addition. Moreover, $\Phi(x, m, n) = x \cdot m + n$. Both laws of distributivity are valid.

$\mathfrak{X}(D_{11})$ is a division ring (multiplication is associative).

$\mathfrak{X}(P_{10})$ is a field.

## § 3.3. Some complementary results.

### Change of coordinate system.

The ternary fields which we obtain from a projective plane $\mathfrak{P}_0$, starting with different coordinate systems $OXYE$ and $O'X'Y'E'$, need not be isomorphic. We shall prove that they are isomorphic in $\mathfrak{P}(D_{11})$. Actually, this is already true in $\mathfrak{P}(D_{10})$, but we shall not prove this theorem. [See e.g. Pickert, Projektive Ebenen § 7.3.]

---

[1]) This proof was suggested to me by E. R. Paerl.

**Theorem 3.3.1.** In $\mathfrak{P}(D_{11})$ the ternary fields attached to $\mathfrak{P}_0(D_{11})$ by different coordinate systems, are isomorphic.

PROOF. Let $\mathfrak{T}$ correspond to $OXYE$, $\mathfrak{T}'$ to $O'X'Y'E'$, $OE \cap XY = W$, $O'E' \cap X'Y' = W'$. As $D_{11}$ is valid, $\mathfrak{T}$ is completely determined by its addition and multiplication (Th. 3.2.4). We have seen in the proofs of Th. 3.2.1 and Th. 3.2.2 that $x + z = y$ is equivalent to $(OzW^2)(OxW^2) = (OyW^2)$ and $x \cdot z = t$ is equivalent to $(EzO^2W^2)(ExO^2W^2) = (EtO^2W^2)$. By Th. 2.5.1, the second order projectivities are independent of $X$ and $Y$; analogously for $X'$ and $Y'$.

Let us suppose that $OE \neq O'E'$ and that none of the points $E$, $W$, $E'$, $W'$ coincides with $OE \cap O'E'$. Put $EE' \cap WW' = S_1$; let $l''$ be a line through $E'$ ($l'' \neq O'E'$). By projecting $O$, $E$, $W$ from $S_1$ on $l''$, we obtain $O''$, $E'$, $W''$. Put $O'O'' \cap W'W'' = S_2$. By projecting $O''$, $E'$, $W''$ from $S_2$ on $O'E'$, we obtain $O'$, $E'$, $W'$. Hence there is a projectivity that transforms $O$, $E$, $W$ into $O'$, $E'$, $W'$ respectively. From Cor. 2.5.7 we see that $(OzW^2)(OxW^2) = (OyW^2)$ will be equivalent to $(O'z'W'^2)(O'x'W'^2) = (O'y'W'^2)$ and $(EzO^2W^2)(ExO^2W^2) = (EtO^2W^2)$ will be equivalent to $(E'z'O'^2W'^2)(E'x'O'^2W'^2) = (E't'O'^2W'^2)$. In other words $x+z=y$ if and only if $x' + z' = y'$; $x \cdot z = t$ if and only if $x' \cdot z' = t'$. This means that the ternary fields are isomorphic.

**Exercise.** The reader may examine the special cases which were excluded here; projection onto an auxiliary line reduces them to the general case.

### Coordinates and duality.

Together with a projective plane $\mathfrak{P}_0 = \langle \Pi, \Lambda, I \rangle$ we may consider the dual plane $\langle \Lambda, \Pi, I \rangle = \mathfrak{P}_D$. In $\mathfrak{P}_D$ we can introduce coordinates by the methods described above. We use the $\mathfrak{P}_0$-terminology (that is, we call an element of $\Lambda$ a line, an element of $\Pi$ a point). In $\mathfrak{P}_0$ a line $l$ has an equation $y = \Phi(x, m, n)$; in $\mathfrak{P}_D$ it has coordinates $(\xi, \eta)$ in a ternary field $\mathfrak{T}_D$. Assuming $D_{10}$, we shall show that the coordinate system in $\mathfrak{P}_D$ can be chosen in such a way that $\mathfrak{T}_D$ is anti-isomorphic to $\mathfrak{T}_0$, and that by this anti-isomorphism the coordinates of the line $y=xm-n$ correspond to $(m, n)$.

Let us take the Cartesian plane as an example to see how the coordinate system $O_D X_D Y_D E_D$ must be chosen. Here $m = 0$ for lines through $X$, so $O_D Y_D = X$. $n = 0$ for lines through $O$, so $O_D X_D = O$. Lines through $Y$ have no pairs of coordinates, so $X_D Y_D = Y$. We must take $O_D = OX$, $X_D = OY$, $Y_D = XY$. $E_D$ must be the line $y = x - 1$; that is the line through $W$ and the point $Q = O_D E_D = (1, 0)$ on $OX$. It follows that $W_D = YE$.

**Theorem 3.3.2.** In $\mathfrak{P}(D_{10})$: Choose in $\mathfrak{P}_D$ the coordinate system $O_D X_D Y_D E_D$ with $O_D = OX$, $X_D = OY$, $Y_D = XY$; $E_D$ is the line $y = x - 1$. The elements of $\mathfrak{X}_D$ are the lines with equations

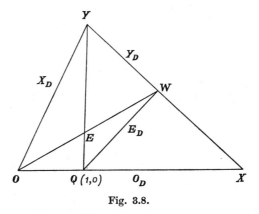

Fig. 3.8.

$y = xm - m$. The mapping of $\mathfrak{X}_D$ onto $\mathfrak{X}_0$ in which the line $y = xm - m$ corresponds to $m$, is an anti-isomorphism. If we identify corresponding elements in this mapping, then the coordinates of $y = xm - n$ are $(m, n)$.

**PROOF.** Let us denote addition and multiplication in $\mathfrak{X}_D$ by $+_D$ and $\cdot_D$ respectively. Denote the line $y = xm - m$, considered as a point in $\mathfrak{P}_D$, by $\overline{m}$, considered as a line in $\mathfrak{P}_0$, by $\underline{m}$.

By the definition of the sum, § 3.2,(9), we have

$$\overline{m} +_D \overline{n} = [(X_D \overline{n} \cap O_D Y_D) W_D \cap Y_D \overline{m}] X_D \cap O_D E_D =$$
$$= [\{(OY \cap \underline{n}) X \cap YE\}(XY \cap \underline{m}) \cap OY] Q,$$

where $Q = (1, 0)$. The equation of $\underline{n}$ is $y = xn - n$. $OY \cap \underline{n}$ is $A = (0, -n)$. $AX$ has the equation $y = -n$; $YE$ has the equation

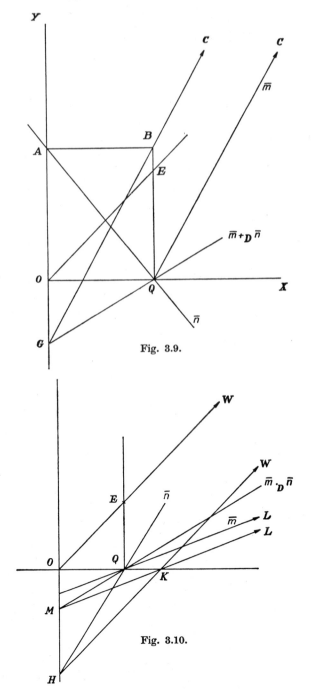

Fig. 3.9.

Fig. 3.10.

$x = 1$, thus $AX \cap YE$ is $B = (1, -n)$. The equation of $m$ is $y = xm - m$. Any line through $C = XY \cap m$ has an equation of the form $y = xm + p$. For $BC$ we have

$$-n = 1 \cdot m + p; \text{ thus } p = -m - n.$$

So the equation of $BC$ is $y = xm - m - n$. Its intersection $G$ with $OY$ is $(0, -m-n)$. $QG$ has the equation $y = x(m+n) - (m+n)$. As $QG$ is $\bar{m} + {}_D\bar{n}$, we have proved that $\bar{m} + {}_D\bar{n}$ is mapped on $m + n$.

By the definition of the product, § 3.2,(10), we have

$$\bar{m} \cdot_D \bar{n} = [(X_D\bar{n} \cap Y_D E_D) O_D \cap Y_D \overline{m}] X_D \cap O_D E_D =$$
$$= [\{(OY \cap \underline{n})W \cap OX\}(XY \cap \underline{m}) \cap OY]Q.$$

$OY \cap \underline{n}$ is $H = (0, -n)$. $HW$ has the equation $y = x - n$. $HW \cap OX$ is $K = (n, 0)$. $XY \cap \underline{m} = L$; any line through $L$ has an equation of the form $y = xm + q$. For $KL$ we have

$$0 = nm + q; \text{ thus } q = -nm.$$

Hence the equation of $KL$ is $y = xm - nm$. $KL \cap OY$ is $M = (0, -nm)$. $QM$ is $y = x(nm) - nm$. As $QM$ is $\bar{m} \cdot_D \bar{n}$, we have proved that $\bar{m} \cdot_D \bar{n}$ is mapped on $nm$. This completes the proof of Th. 3.3.2.

## Algebraic consequences of $D_9$.

**Theorem 3.3.3.** In $\mathfrak{P}(D_9)$: If $ab = 1$, then $ba = 1$ and, for every $x$, $(xa)b = b(ax) = x$.

PROOF. The construction of $ab$, given in § 3.2 (10), together with the fact that $ab = 1$, leads to the following figure. $Xb \cap YE = P$, $OP \cap XE = Q$, $YQ \cap OE = a$. $ba$ is constructed as follows: $Xa \cap YE = R$, $OR \cap Yb = S$, $XS \cap OE = ba$. It must be proved that $ba = 1$, or, what comes to the same, that $QS$ contains $X$. This is a consequence of $P_9$ applied to the hexagon $bPQaRS'$, where $S' = QE \cap OR$. As $P_9$ holds in $\mathfrak{P}(D_9)$ (Th. 2.6.9), the first part of the theorem has been proved.

In the rest of the proof we use the theory of harmonic pairs, which has been developed in $\mathfrak{P}(D_9)$; see theorems 2.4.2—2.4.5.

Denote $PS \cap OW$ by $H$, $PS \cap XY$ by $K$. The quadrangle $EbXY$ shows that $HK$ harm. $PS$; the quadrangle $EbPS$ shows

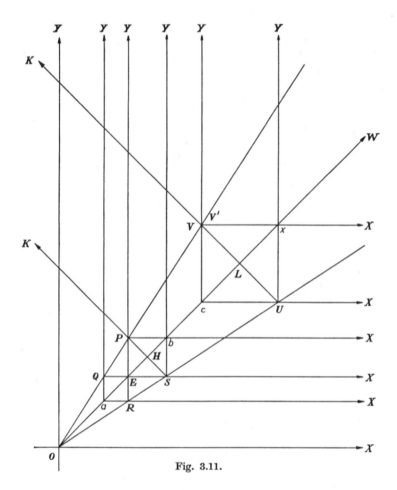

Fig. 3.11.

that $WK$ harm. $XY$, so that $K$ does not depend upon the choice of $b$.

Now let $x$ be any point on $OE$. $(xa)b$ is constructed as follows. $OR \cap Yx = U$, $UX \cap OE = c$, $(c = xa)$, $OQ \cap Yc = V$, $VX \cap OE = (xa)b$. It must be proved that $VX$ contains $x$.

Denote $Xx \cap Yc$ by $V'$; $UV' \cap OW$ by $L$. The quadrangle $cxUV'$ shows that $V'U$ intersects $XY$ in the harmonic conjugate of $W$ with respect to $X$, $Y$; that is in $K$. The quadrangle $cxXY$ shows that $LK$ harm. $UV'$. But, by projection of the harmonic pairs $P$, $S$ and $H$, $K$ from $O$, we see that $OQ$ intersects $UV'$ in the harmonic conjugate of $U$ with respect to $L$, $K$; thus $V' \in OQ$, $V' = V$. This proves that $(xa)b = x$.

An analogous proof can be given for $b(ax) = x$, but in $\mathfrak{P}(D_{10})$ it is easier to apply the duality principle, which holds in $\mathfrak{P}(D_{10})$ (Th. 2.2.7). As in the proof of Th. 3.3.2, let us denote the line $y = xm-m$ by $\bar{m}$, when considered as a point in $\mathfrak{P}_D$. If $ba = 1$, then $\bar{a} \cdot \bar{b} = \bar{1}$. Applying the preceding result to $\mathfrak{P}_D$, we obtain $(\bar{x} \cdot_D \bar{a}) \cdot_D \bar{b} = \bar{x}$, so $b \cdot (a \cdot x) = x$.

## § 3.4. The geometry over a given ternary field.

Let $\mathfrak{T}$ be any ternary field, defined by the function $\Phi(x, m, n)$. In order to construct a projective plane over $\mathfrak{T}$, we define the sets $\Pi_0$ and $\Lambda_0$ and the incidence relation $I_0$.

$\Pi_0$ is the union of $\Pi_1$, $\Pi_2$, $\Pi_3$; $\Lambda_0$ is the union of $\Lambda_1$, $\Lambda_2$, $\Lambda_3$.
$\Pi_1$ is the set of pairs $(x, y)$ of elements of $\mathfrak{T}$.
$\Pi_2$ is the set of elements $(p)$ of $\mathfrak{T}$.
$\Pi_3$ consists of one element $Y$, not in $\Pi_1 \cup \Pi_2$.
$\Lambda_1$ is the set of pairs $[m, n]$ of elements of $\mathfrak{T}$.
$\Lambda_2$ is the set of elements $[c]$ of $\mathfrak{T}$.
$\Lambda_3$ consists of one element $\omega$, not in $\Lambda_1 \cup \Lambda_2 \cup \Pi_0$.

REMARK. In order to make $\Pi_1$ and $\Lambda_1$ into disjoint sets, we might introduce elements $\alpha$, $\beta$ not belonging to $\mathfrak{T}$, and take the triples $(x, y, \alpha)$ as the elements of $\Pi_1$, $(m, n, \beta)$ as the elements of $\Lambda_1$; analogously for $\Pi_2$ and $\Lambda_2$. Here I use parentheses ( ) and square brackets [ ] instead of $\alpha$ and $\beta$.

The relation $I_0$ is defined as follows.

$$(*) \begin{cases} (x, y) \; I_0 \, [m, n] & \text{if } y = \Phi(x, m, n). \\ (x, y) \; I_0 \, [c] & \text{if } x = c. \\ (p) \qquad I_0 \, [m, n] & \text{if } p = m. \\ (p) \qquad I_0 \, \omega & \text{for every } p. \\ Y \qquad I_0 \, [c] & \text{for every } c. \\ Y \qquad I_0 \, \omega. \\ \text{No other incidences occur.} \end{cases}$$

This triple $\langle \Pi_0, \Lambda_0, I_0 \rangle$ satisfies axioms V1, V2, V3; thus it forms a projective plane which we shall denote by $\mathfrak{P}(\mathfrak{T})$. For every ternary field $\mathfrak{T}$, $\mathfrak{P}(\mathfrak{T})$ is a model of V1, V2, V3.

**Exercise.** Verify the axioms V1, V2, V3 for $\mathfrak{P}(\mathfrak{T})$ by using the defining properties of a ternary field.

REMARK. If $\mathfrak{T}$ is the field of real numbers, $\mathfrak{P}(\mathfrak{T})$ becomes the Euclidean plane, extended with the points at infinity and with the line at infinity. After the introduction of homogeneous coordinates, this plane becomes isomorphic to $APPG$.

If we start with a projective plane $\mathfrak{P}_0$ and a coordinate system $OXYE$ in $\mathfrak{P}_0$, these determine a ternary field $\mathfrak{T}_0$, as described in section 3.1. It is easily seen from the definition of $\mathfrak{P}(\mathfrak{T}_0)$, that $\mathfrak{P}(\mathfrak{T}_0) \cong \mathfrak{P}_0$.

Conversely, let us start with a ternary field $\mathfrak{T}_1$. In $\mathfrak{P}(\mathfrak{T}_1)$ we choose the coordinate system $OXYE$ with $O = (0, 0)$, $X = (0)$, $Y = Y$, $E = (1, 1)$. Then $OE = [1, 0]$, and $(x, x) \in OE$ for every $x$ in $\mathfrak{T}_1$. The elements of $OE$ form the ternary field $\mathfrak{T}_2$, attached to $\mathfrak{P}(\mathfrak{T}_1)$ by the coordinate system $OXYE$. In $\mathfrak{T}_2$ the function $\Phi_2$, defined on $OE$ occurs. We shall prove that the correspondence $\sigma$ defined by $\sigma x = (x, x)$ is an isomorphism between $\mathfrak{T}_1$ and $\mathfrak{T}_2$.

If $P = (p, q)$, let us call $p, q$ the analytic coordinates of $P$. The coordinates of $P$ in $OXYE$ are $(\sigma p, \sigma q)$; let us call these the geometric coordinates of $P$. Let $l$ be the line $[m, n]$. Using analytic coordinates, we see from (*) that $Pll$ if and only if $q = \Phi(p, m, n)$.

Moreover, $R = l \cap OY = (0, n)$; $S = l \cap XY = (m)$; $T = OS \cap EY = (1, m)$. Thus, in geometric coordinates, $R = (\sigma 0, \sigma n)$; $T = (\sigma 1, \sigma m)$; $S = OT \cap XY$; $l = RS$; $P = l \cap (Y \sigma p)$; $\sigma q = PX \cap OE$. This is exactly the construction of $\sigma q = \Phi_2(\sigma p, \sigma m, \sigma n)$, according to (8) of section 3.1.

Thus $Pll$ if and only if $\sigma q = \Phi_2(\sigma p, \sigma m, \sigma n)$. We know now that $q = \Phi(p, m, n)$ is equivalent to $\sigma q = \Phi_2(\sigma p, \sigma m, \sigma n)$. This proves that $\sigma$ is an isomorphism.

**Theorem 3.4.1.** To every ternary field $\mathfrak{T}$ there corresponds a projective plane $\mathfrak{P}(\mathfrak{T})$ and a coordinate system $\mathfrak{S}$ in $\mathfrak{P}(\mathfrak{T})$, such that the ternary field attached to $\mathfrak{P}(\mathfrak{T})$ by $\mathfrak{S}$ is isomorphic to $\mathfrak{T}$. Conversely, if the projective plane $\mathfrak{P}_1$, together with the coordinate system $\mathfrak{S}_1$, determine the ternary field $\mathfrak{T}_1$, then $\mathfrak{P}(\mathfrak{T}_1)$ is isomorphic to $\mathfrak{P}_1$.

In the preceding sections we have found that additional axioms, e.g. $D_{10}$, $D_{11}$ or $P_{10}$, involve certain properties of the ternary field which is attached to the plane. We shall now prove that, conversely, additional properties of the ternary field $\mathfrak{T}$ involve the validity of certain propositions in $\mathfrak{P}(\mathfrak{T})$.

Let us first consider $D_{10}$. $\mathfrak{P}(D_{10})$ is called a *Moufang plane*. In $\mathfrak{T}(D_{10})$, the following properties hold:

(i)    $\Phi(x, m, n) = xm + n.$      [Th. 3.2.4]
(ii)   $(x + y) + z = x + (y + z).$    [Th. 3.2.1]
(iii)   $x + y = y + x.$      [Th. 3.2.1]
(iv)   $x(y + z) = xy + xz.$      [Th. 3.2.7]
(v)    $(x + y)z = xz + yz.$      [Th. 3.2.6]
(vi)   $(xy)z = x$ if $yz = 1.$      [Th. 3.3.3]
(vii)   $x(yz) = z$ if $xy = 1.$      [Th. 3.3.3]

We now prove the converse:

**Theorem 3.4.2.** If the equations (i)—(vii) are universally true in $\mathfrak{T}$, then $D_{10}$ is valid in $\mathfrak{P}(\mathfrak{T})$.

PROOF. By Th. 2.3.9, it is sufficient to prove that $\mathfrak{P}(\mathfrak{T})$ is $P$-$\xi$-transitive for every point $P$ and every line $\xi$ with $P \,\epsilon\, \xi$. [1]) In a number of lemmas we shall prove the $P$-$\xi$-transivity for special choices of $P$ and $\xi$.

**Lemma 1.** If, for a given point $A$ $(A \neq P, A \notin \xi)$ and any point $B$ on $PA$ $(B \neq P, B \notin \xi)$ the central collineation $(ABP^2\xi^2)$ exists, then $\mathfrak{P}$ is $P$-$\xi$-transitive.

PROOF. It is easy to find $B$ such that $(ABP^2\xi^2)$ transforms a given point $Q$ into a given point $R$ on $PQ$ $(Q, R \neq P; Q, R \notin \xi)$.

**Lemma 2.** If $\mathfrak{P}$ is $P$-$\xi$-transitive, and $\mathfrak{C}$ is a collineation, then $\mathfrak{P}$ is $\mathfrak{C}P$-$\mathfrak{C}\xi$-transitive.

PROOF. $((\mathfrak{C}A)(\mathfrak{C}B)(\mathfrak{C}P)^2(\mathfrak{C}\xi)^2) = \mathfrak{C}(ABP^2\xi^2)\mathfrak{C}^{-1}$; it is easily verified that the right hand side is a central collineation with centre $\mathfrak{C}P$ and axis $\mathfrak{C}\xi$.

**Lemma 3.** It follows from (i), (ii), (iii) and (iv) that $\mathfrak{P}(\mathfrak{T})$ is $Y$-$OY$-transitive.

---

[1]) For the equivalence of $D_{10}$ and $D_{10}{}^{\text{I}}$, see Th. 2.2.8 and Th. 2.2.10.

PROOF. By lemma 1 it suffices to prove that $(XUY^2\overline{OY^2})$ exists for every point $U$ on $XY$ ($U \neq Y$). Let $P = (p, q)$ be any point outside $XY$. $P'$ is constructed as follows: $XP \cap OY = T$; $TU \cap YP = P'$. $T = (0, q)$. The equation of $TU$ is $y = xm + q$, where $m$ is an element depending on $U$. So $P' = (p', q')$ with $p' = p$, $q' = pm + q$.

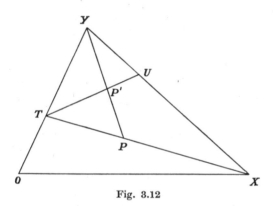

Fig. 3.12

The equations $x' = x$, $y' = xm + y$ define a transformation $\mathfrak{S}$. The line $y = xr + s$ is transformed into $y = x(m + r) + s$. Thus, if we extend $\mathfrak{S}$ to $XY$ by assigning $(m + r)$ to $(r)$, then $\mathfrak{S}$ transforms lines into lines. This proves that $\mathfrak{S}$ is a central collineation with the desired properties.

**Lemma 4.** It follows from (i), (ii), (iii) and (v), that $\mathfrak{P}(\mathfrak{T})$ is $X$-$XY$-transitive.

PROOF. Let $\sigma$ be the transformation $x' = x + a$, $y' = y$. If the equation of $l$ is $y = xm + n$, then that of $\sigma l$ is $y = (x - a)m + n$; thus $\sigma l$ is the line $y = xm + (-am + n)$. If we extend $\sigma$ to $XY$ by fixing that it is the identity on $XY$, then $\sigma$ becomes a central collineation; moreover, if $A = (a, 0)$, then $\sigma = (OAX^2\overline{XY^2})$. This proves lemma 4.

**Lemma 5.** It follows from (i) - (vii) that $\mathfrak{P}(\mathfrak{T})$ is $X - OX -$ transitive.

PROOF. Let $A$ be the point $(a)$ on $XY$. We prove the existence of $(YAX^2\overline{OX^2})$. If this central collineation exists, then the image $P'$ of a point $P$ outside $XY$ is constructed as follows:

$P' = (YP \cap OX) A \cap XP$. If $P = (p_1, p_2)$, then $B = YP \cap OX = (p_1, 0)$. It is easily seen that the equation of $AB$ is $y = xa - p_1 a$; thus, if $P' = (p_1', p_2)$, then $p_2 = p_1' a - p_1 a$; $p_1' = p_1 + p_2 a^{-1}$.

Let $\sigma$ be the mapping with equations $x' = x + ya^{-1}$, $y' = y$; we must show that $\sigma$ is a central collineation. $\sigma$ maps the line $y = xm + n$ on $y = (x - ya^{-1})m + n$; this can be successively reduced to $ym^{-1} = x - ya^{-1} + nm^{-1}$, $y(m^{-1} + a^{-1}) = x + nm^{-1}$; $y = x(m^{-1} + a^{-1})^{-1} + (nm^{-1})(m^{-1} + a^{-1})^{-1}$, which is again the equation of a line. It is clear that $\sigma$ leaves every line through $X$ and every point on $OX$ invariant. This proves lemma 5. The reader may verify that in the calculations above only (i)—(vii) and the general properties of a ternary field are used. He may also treat the case that $m = -a$.

**Lemma 6.** It follows from (i)—(vii) that $\mathfrak{P}(\mathfrak{T})$ is $O$–$OY$-transitive.

PROOF. In the proof of Th. 3.3.2. only the properties (i)—(v) and the general properties of a ternary field were used; therefore we can infer from this proof: If (i)—(v) hold in $\mathfrak{T}(\mathfrak{P})$, then $\mathfrak{T}(\mathfrak{P}_D)$ is anti-isomorphic to $\mathfrak{T}(\mathfrak{P})$. Consequently, if (i)—(vii) hold in $\mathfrak{T}(\mathfrak{P})$, then they hold in $\mathfrak{T}(\mathfrak{P}_D)$. Then, by lemma 5, $\mathfrak{P}_D$ is $X_D$–$O_D X_D$-transitive; this is the same as saying that $\mathfrak{P}$ is $O$–$OY$-transitive.

**Lemma 7.** Let $P$ be any point not on $OX$ or $OY$, and $\xi$ a line through $P$, not through $Y$, then it follows from (i)—(vii) that $\mathfrak{P}(\mathfrak{T})$ is $P$-$\xi$-transitive.

PROOF. Put $OP \cap XY = Q, \xi \cap OY = R, QR = \eta, \eta \cap OX = T$,

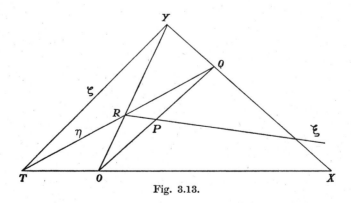

Fig. 3.13.

$YT = \zeta$. By lemma 3, $\mathfrak{P}$ is $Y$-$OY$-transitive. By lemma 4, $\mathfrak{C}_1 = (OTX^2\overline{XY}^2)$ exists; $\mathfrak{C}_1 Y = Y$, $\mathfrak{C}_1(OY) = \zeta$. Thus, by lemma 2, $\mathfrak{P}$ is $Y$-$\zeta$-transitive.

By lemma 5, $\mathfrak{C}_2 = (YQX^2\overline{OX}^2)$ exists; $\mathfrak{C}_2 Y = Q$, $\mathfrak{C}_2\zeta = \eta$. Thus, by lemma 2, $\mathfrak{P}$ is $Q$-$\eta$-transitive. By lemma 6, $\mathfrak{C}_3 = (QPO^2\overline{OY}^2)$ exists; $\mathfrak{C}_3 Q = P$, $\mathfrak{C}_3\eta = \xi$. Thus, again by lemma 2, $\mathfrak{P}$ is $P$-$\xi$-transitive.

**Lemma 8.** In the cases where $P \in OX$ or $P \in OY$ or $Y \in \xi$, $\mathfrak{P}(\mathfrak{T})$ is $P$-$\xi$-transitive for $P \in \xi$.

PROOF. We shall only prove the case $P \in OX$, $P \notin OY$, $Y \notin \xi$; the reader may prove the other cases in a similar way. Put $\xi \cap OY = R$, choose $T \in PY$ ($T \neq P, Y$). By lemma 7, $\mathfrak{P}$ is $T$-$TR$-transitive; if $\mathfrak{C}_4 = (TPY^2\overline{OY}^2)$, then $\mathfrak{C}_4 T = P$, $\mathfrak{C}_4(TR) = \xi$. Thus, by lemma 2, $\mathfrak{P}$ is $P$-$\xi$-transitive.

This completes the proof of Th. 3.4.2.

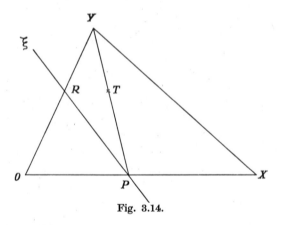

Fig. 3.14.

REMARK. A ternary field in which (i)—(vii) hold, is called an *alternative field*. Hence $\mathfrak{T}(\mathfrak{P}_0)$ is an alternative field if an only if $D_{10}$ holds in $\mathfrak{P}_0$. The algebraic theory of alternative fields leads to interesting results; see e.g., Pickert, Projektive Ebenen chapter 6 and the literature mentioned there.

**Theorem 3.4.3.** If $\mathfrak{T}$ is a division ring, then $D_{11}$ is valid in $\mathfrak{P}(\mathfrak{T})$.

This can be proved by a method analogous to that used in the preceding theorem. It is also possible to prove the theorem directly by the methods of analytic geometry; the calculations in this proof are complicated, because we have not proved that, if $\mathfrak{T}$ is a division ring in one coordinate system, it is also a division ring in every other coordinate system. Therefore no special coordinate system can be used in the proof. The isomorphy of the ternary fields for all coordinate systems, if one of them is a division ring, then follows from Th. 3.3.1.

Another proof of Th. 3.4.3 will be given in Chapter V; see the remark after Th. 5.4.1.

REMARK. If a ternary field satisfies (i)—(vii) of Th. 3.4.2 and the associative law of multiplication, then it is a division ring. Thus, if $D_{10}$ is valid in $\mathfrak{P}(\mathfrak{T})$ and multiplication is associative in $\mathfrak{T}$, then $D_{11}$ is valid in $\mathfrak{P}(\mathfrak{T})$.

**Theorem 3.4.4.** If $\mathfrak{T}$ is a field, then $P_{10}$ is valid in $\mathfrak{P}(\mathfrak{T})$.

PROOF. Any proof of Pappos' theorem in analytic projective geometry can be used here. We may even give a proof in which a special coordinate system is introduced, for by Th. 3.4.3, $D_{11}$ is valid, so by Th. 3.3.1 the ternary fields corresponding to different coordinate systems are all isomorphic.

REMARK. It follows from Th. 3.2.8 and Th. 3.4.4:

If $D_{11}$ is valid in $\mathfrak{P}(\mathfrak{T})$ and multiplication is commutative in $\mathfrak{T}$, then $P_{10}$ is valid in $\mathfrak{P}(\mathfrak{T})$.

## § 3.5. Independence results.

The only independence results proved so far are Th. 2.2.1 and the stronger Th, 2,2,13, We can now prove

**Theorem 3.5.1.** $D_{11}$ is independent of $V_1$, $V_2$, $V_3$, $D_{10}$.

PROOF. Cayley's algebra (§ 1.3.3), is a ternary field if $\Phi$ is defined by (i) of Th. 3.4.2; (ii)–(vii) of Th. 3.4.2 hold, but multiplication is not associative. Thus in $\mathfrak{P}(\mathfrak{C})$, $D_{10}$ is a theorem, but $D_{11}$ is not valid in view of Th. 3.2.2.

**Theorem 3.5.2.** $P_{10}$ is independent of $V_1$, $V_2$, $V_3$, $D_{11}$.

PROOF. As real quaternions form a non-commutative division ring $\mathfrak{Q}$ (§ 1.3.2), $D_{11}$ holds in $\mathfrak{P}(\mathfrak{Q})$, (Th. 3.4.3), but $P_{10}$ is not valid (Th. 3.2.3).

## § 3.6. Homogeneous coordinates.

So far, coordinates were only introduced in the affine plane; points on $XY$ had no coordinates. In the case of $\mathfrak{P}(D_{11})$, where the coordinates belong to a division ring, the assignment of coordinates can be extended to the whole projective plane by the use of homogeneous coordinates, analogous to those which were used in § 1.4 for $APPG$.

The first thing to do is to extend the notion of ratio to division rings.

**Definition.** Two sequences $(x_1, \ldots, x_n)$ and $(y_1, \ldots, y_n)$ of elements of a division ring $\mathfrak{F}$ are called *left-proportional* if there exists an element $c \neq 0$ of $\mathfrak{F}$ such that $y_i = cx_i$ $(i = 1, \ldots, n)$.

It is easily seen that left-proportionality is an equivalence relation in $\mathfrak{F}^n$.

**Definition.** An equivalence class in $\mathfrak{F}^n$ with respect to the relation of left-proportionality, which does not consist of the null sequence, is called a *left n-ratio*. (Where the number $n$ is clear from the context, we simply speak of a left ratio.)

The definitions of *right-proportionality* and of *right ratio* are analogous.

Now let us consider a projective plane $\mathfrak{P}_0(D_{11})$. Let $(x, y)$ be the coordinates of a point $P$ in some coordinate system $OXYE$. Any triple $(x_0, x_1, x_2)$ in the left ratio $(1, x, y)$ is a triple of *homogeneous coordinates* for $P$. In other words, if $x_0 \neq 0$, then a triple of homogeneous coordinates for $P$ is $(x_0, x_0 x, x_0 y)$.

Analogously, let $y = xm + n$ be the equation of a line $l$. Any triple $(\xi_0, \xi_1, \xi_2)$ in the right ratio $(n, m, -1)$ is called a triple of homogeneous coordinates for $l$. Thus, if $\xi_2 \neq 0$, then $(-n\xi_2, -m\xi_2, \xi_2)$ is a triple of homogeneous coordinates for $l$. The equation of $l$ can be written as

$$n + xm - y = 0, \qquad \text{or as}$$

(1) $$x_0(n + xm - y)\xi_2 = 0 \quad \text{or as}$$

$$x_0\xi_0 + x_1\xi_1 + x_2\xi_2 = 0.$$

It is here that the associative rule for multiplication is used. It is also at this point that it becomes clear why left ratios are used for the coordinates of points and right ratios for those of lines.

Our aim is to extend the use of homogeneous coordinates to the points on $XY$ and to the lines through $Y$, which until now where excluded. Let $Q$ be a point on $XY$, $Q \neq Y$, and let $y = xm$ be the equation of $OQ$. Every line $s$ through $Q$ (except $XY$) has an equation $y = xm + n$ and homogeneous coordinates $(n, m, -1)$ If we assign to $Q$ the homogeneous coordinates $(0, 1, m)$, then (1) will be satisfied by $Q$ and $s$. A line $t$ through $Y$ has an equation $x = c$; a point $R$ on $t$ has coordinates $(1, c, d)$. If we assign to $t$ the homogeneous coordinates $(-c, 1, 0)$, then (1) is again satisfied by $R$ and $t$. Finally, we give $Y$ the coordinates $(0, 0, 1)$ and $XY$ the coordinates $(1, 0, 0)$; then (1) is a necessary and sufficient condition for the point $(x_0, x_1, x_2)$ to be incident with the line $(\xi_0, \xi_1, \xi_2)$.

**Theorem 3.6.1.** In every projective plane $\mathfrak{P}_0(D_{11})$, we can assign left-homogeneous coordinates $(x_0, x_1, x_2)$ to the points and right-homogeneous coordinates $(\xi_0, \xi_1, \xi_2)$ to the lines, such that (1) is the condition for incidence of point and line.

From Th. 3.6.1 the duality of coordinates is evident. For let $\mathfrak{F}^*$ be the division ring which has the same elements as $\mathfrak{F}$, but with the order of multiplicands in a product reversed; that is, if $ab = c$ in $\mathfrak{F}$, then $ba = c$ in $\mathfrak{F}^*$. (It is easy to see that $\mathfrak{F}^*$ is again a division ring.) If we map the point $(x_0, x_1, x_2)$ of $\mathfrak{P}(\mathfrak{F})$ on the line $(x_0, x_1, x_2)$ of $\mathfrak{P}(\mathfrak{F}^*)$ and the line $(\xi_0, \xi_1, \xi_2)$ of $\mathfrak{P}(\mathfrak{F})$ on the point $(\xi_0, \xi_1, \xi_2)$ of $\mathfrak{P}(\mathfrak{F}^*)$, then the incidence relations will be preserved. Thus $\mathfrak{P}(\mathfrak{F}^*)$ is the dual of $\mathfrak{P}(\mathfrak{F})$ in the sense of § 2.1 (see the remark after Th. 2.1.3).

A corresponding result in the affine plane was proved in Th. 3.3.2 under the weaker assumption $D_{10}$.

# CHAPTER IV

# INCIDENCE PROPOSITIONS IN SPACE

## § 4.1. Trivial axioms and propositions.

**Definition.** A three-dimensional projective geometry $\mathfrak{P}^3$(R1—R5)[1] is an axiomatic theory with as set of fundamental notions the quadruple $\langle \Pi, \Lambda, \Sigma, I \rangle$ and as axioms R1—R5 below. $\Pi$, $\Lambda$, $\Sigma$ are disjoint sets; the elements of $\Pi$ are called points, those of $\Lambda$ lines, those of $\Sigma$ planes. $I$ is a symmetric relation which can exist between a point and a line, between a point and a plane and between a line and a plane. $aIb$ is to be read "$a$ is incident with $b$".

## Axioms.

**R1.** Given two different points, there is exactly one line with which both are incident.

**R2.** Given a point and a line, which are not incident, there is exactly one plane with which both are incident.

**R3.** Given a plane and a line, which are not incident, there is exactly one point with which both are incident.

**R4.** If the point $P$ is incident with the line $l$, and $l$ is incident with the plane $\alpha$, then $P$ is incident with $\alpha$.

**R5.** There exist five points $A_1$, $A_2$, $A_3$, $A_4$, $A_5$, which are not all incident with one line, and no four of which are incident with a plane.

**Exercise.** The reader may verify that R1—R5 are valid in $ASPG$ (§ 1.5).

**REMARKS.** $A$, $B$, $P$, $Q$, ... will denote points, $l$, $m$, $p$, ..., lines, $\alpha$, $\beta$, $\gamma$, ..., planes. The line incident with $A$ and $B$ will be denoted by $AB$, the plane incident with $P$ and $l$ by $Pl$, the point incident with $\alpha$ and $l$ by $\alpha \cap l$.

---

[1] The symbols $\mathfrak{P}^2$ and $\mathfrak{P}^3$ will denote a two- and a three-dimensional projective geometry respectively.

The following expressions will be used synonymously:

$PIl$, $lIP$, $P$ lies on $l$, $l$ passes through $P$, $l$ contains $P$.

$PI\alpha$, $\alpha IP$, $P$ lies in $\alpha$, $\alpha$ passes through $P$, $\alpha$ contains $P$.

$lI\alpha$, $\alpha Il$, $l$ lies in $\alpha$, $\alpha$ passes through $l$, $\alpha$ contains $l$.

**Exercise.** In R5 we postulate that the five points are not collinear, in order to avoid the following trivial model: $\Pi$ consists of five points, all incident with a line $l$ that is the only element of $\Lambda$; $\Sigma$ contains no element at all. Verify the axioms for this model.

**Remark.** As in Chapter 2, $\mathfrak{P}^3$ denotes an axiomatic theory, i.e. a set of theorems. Models of $\mathfrak{P}^3$ will be denoted by $\mathfrak{P}_0^3$, $\mathfrak{P}_1^3$,...; such models are called projective spaces. Moreover, $\mathfrak{P}_0^3(A_1,...,A_k)$ denotes a projective space in which the axioms $A_1$, , ..., $A_k$ are valid.

**Exercise.** The proofs of the following theorems are not worked out in detail. It is left to the reader to complete them and to examine step by step that the reasoning follows from the axioms without use of space-intuition.

**Theorem 4.1.1.** In $\mathfrak{P}^3$ (R1—R5): A line which contains two different points of a plane, lies in that plane.

Proof. R3.

**Theorem 4.1.2.** In $\mathfrak{P}^3$ (R1—R5): Given three points which are not on a line, there is exactly one plane which is incident with these points.

Proof. Easy; use R1, R2, R4. Th. 4.1.1.

**Remark.** The plane through $A, B, C$ will be denoted by $ABC$.

**Theorem 4.1.3.** In $\mathfrak{P}^3$ (R1—R5): Every plane contains four points, of which no three are collinear.

Proof. Let $\alpha$ be a plane. Of the five points which exist by R5, at least one, say $A_5$, is not in $\alpha$. The lines $A_iA_5$ ($i = 1, 2, 3, 4$) intersect $\alpha$ in $B_1, B_2, B_3, B_4$. It is easily seen that these points satisfy the condition of the theorem.

**Theorem 4.1.4.** In $\mathfrak{P}^3$ (R1—R5): Given two different planes, there is exactly one line which is incident with both.

REMARK. The line incident with $\alpha$ and $\beta$ will be denoted by $\alpha \cap \beta$.

PROOF. Let $\alpha$, $\beta$ be different planes. Choose $P$, $Q$, $R$ in $\alpha$, not on one line (Th. 4.1.3). These points are not all in $\beta$ (Th. 4.1.2); let $P$ be outside $\beta$. Then $PQ$ and $PR$ intersect $\beta$ in $S$ and $T$ respectively. $S \neq T$, for otherwise $P$, $Q$, $R$ would be collinear. The line $ST = p$ is in $\alpha$ and in $\beta$. If another line $q$ were in $\alpha$ and in $\beta$, then $S$ or $T$ would be outside $q$, so $\alpha$ would coincide with $\beta$.

**Theorem 4.1.5.** In $\mathfrak{P}^3$ (R1—R5), there exist five planes which are not all incident with one line and no four of which are incident with a point.

PROOF. Let $A_1, \ldots, A_5$ be five points satisfying R5. The planes $\alpha_{ikl} = A_i A_k A_l$, where $i$, $k$, $l$ are successively; 1, 2, 3; 2, 3, 4; 3, 4, 5; 4, 5, 1 and 5, 1, 2, satisfy the conditions.

As in the case of the plane, there is a duality principle in space.

**Theorem 4.1.6. (duality principle).** If in a theorem of $\mathfrak{P}^3$ (R1—R5) we interchange the words "point" and "plane", we obtain again a theorem of $\mathfrak{P}^3$ (R1—R5).

PROOF. If in R1—R5 we interchange the words "point" and "plane", we obtain Th. 4.1.4, R3, R2, R4, Th. 4.1.5 respectively. We can now reason exactly as in the proof of Th. 2.1.3.

Thus, if $\langle \Pi_0, \Lambda_0, \Sigma_0, I_0 \rangle$ is a projective space $\mathfrak{P}_0^3$ (R1—R5), then $\langle \Sigma_0, \Lambda_0, \Pi_0, I_0 \rangle$ is also a projective space $\mathfrak{P}_D^3$ (R1—R5). These two spaces are *dual* to each other.

**Theorem 4.1.7.** In $\mathfrak{P}^3$ (R1—R5), the set of points in a plane $\alpha$ and the set of lines in $\alpha$, together with the restriction of $I$ to these sets, form a projective plane $\mathfrak{P}^2(D_{11})$. In other words, every plane in projective space is a projective plane in the sense of section 2.1, and moreover Desargues' proposition holds in it.

PROOF. We must prove that V1, V2, V3, $D_{11}$ hold.

The proof of V1 and V3 is immediate from R1, Th. 4.1.1, Th.4.1.3.

Proof of V2. Let $l$ and $m$ be lines in $\alpha$, $l \neq m$. Choose $A$ outside $\alpha$; $Al = \beta$. $m$ does not lie in $\beta$; $m \cap \beta = S$. If $S$ were not on $l$, then $\alpha$ and $\beta$ would coincide (R2). Thus $l$ intersects $m$ in $S$.

Proof of $D_{11}$. Let the triangles $A_1 A_2 A_3$ and $B_1 B_2 B_3$ in $\alpha$ with the point $O$ satisfy the hypothesis of $D_{11}$ (corresponding vertices are different, corresponding sides are different, lines connecting corresponding vertices pass through $O$). $a_i \cap b_i = C_i$; $C_1 C_2 = l$.

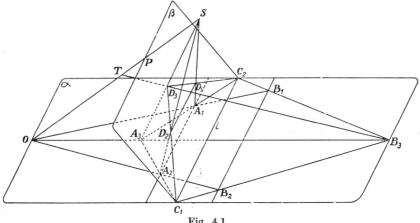

Fig. 4.1.

Choose $P$ outside $\alpha$; $Pl = \beta$. Choose $S$ on $OP$ outside $\alpha$ and $\beta$ (Exercise: show that this is possible). $SA_i \cap \beta = D_i$ $(i = 1, 2, 3)$; the sides of the triangle $D_1 D_2 D_3$ are $d_1$, $d_2$, $d_3$. It is easily seen that $C_1 \in d_1$, $C_2 \in d_2$ and $a_3 \cap l \in d_3$, If $B_3 D_3 \cap OS = T$, we see that $T \in D_1 B_1$ and $T \in D_2 B_2$. It follows that $b_3 \cap l \in d_3$, and thus $C_3 \in l$.

**REMARK.** Denoting $D_1$, $D_2$, $D_3$, $S$, $T$ by $P_1$, $P_2$, $P_3$, $P_4$, $P_5$ respectively, and mapping the intersection of $P_i P_k$ with $\alpha$ on the pair $(i, k)$ of numbers we get exactly the mapping discussed in § 2.2 (see after Th. 2.2.4). Thus Desargues' configuration is the plane intersection of the figure consisting of the lines and planes connecting 5 points in space.

## § 4.2. The Sixteen Points Proposition.

In analytic solid projective geometry $ASPG$, the following proposition is valid.

**Sixteen Points Proposition $R_{16}$.** Let $a_0$, $a_1$, $a_2$, $a_3$, $b_0$, $b_1$, $b_2$, $b_3$ be different lines, not in one plane. If fifteen of the intersections $S_{ik} = a_i \cap b_k$ exist, then the sixteenth exists as well.

In $ASPG$ this theorem is a simple corollary of the theory of lines on a quadric. In the axiomatic theory we have the following theorem.

**Theorem 4.2.1.** In $\mathfrak{P}^3$ (R1—R5), the sixteen points proposition $R_{16}$ is equivalent to Pappos' proposition $P_{10}$.

PROOF.     I. $P_{10} \rightarrow R_{16}$.

Let every $S_{ik}$ be known to exist, except $S_{33}$. Choose a plane $\tau$ which does not contain $S_{00}$, and project the points and lines of the figure from $S_{00}$ on $\tau$. Let $T_{ik}, a_i', b_i'$ be the projections of $S_{ik}$, $a_i$, $b_i$ respectively ($i$, $k = 1, 2, 3$). $a_0 \cap \tau = A_0$, $b_0 \cap \tau = B_0$. Then $a_1'$, $a_2'$, $a_3'$ contain $B_0$; $b_1'$, $b_2'$, $b_3'$ contain $A_0$.

The sides of the hexagon $T_{12}T_{13}T_{23}T_{21}T_{31}T_{32}$ are $a_1'$, $b_3'$, $a_2'$, $b_1'$, $a_3'$, $b_2'$; they pass alternately through $B_0$ and $A_0$. By the dual of $P_{10}$ (see Th. 2.6.8) $T_{12}T_{21}$, $T_{13}T_{31}$, $T_{23}T_{32}$ have a point $R$ in common. Now consider the points $R_1 = S_{12}S_{21} \cap S_{13}S_{31}$, $R_2 = S_{12}S_{21} \cap S_{23}S_{32}$ (these points exist!). Suppose $R_1 \neq R_2$. The projections $R_1'$ and $R_2'$ coincide with $R$. Thus $S_{00}$, $R_1$, $R_2$, $S_{12}$, $S_{21}$ are on one line $l$; it is easily seen that the plane $lS_{01}$ contains all the $a_i$ and $b_i$, which contradicts the hypothesis. It follows that $R_1 = R_2$. Then $S_{13}$, $S_{31}$, $S_{23}$, $S_{32}$ are in a plane, so that $S_{33} = a_3 \cap b_3$ exists.

II. $R_{16} \rightarrow P_{10}$.

We shall prove the dual of $P_{10}$. The proof consists simply in reconstructing the figure of the preceding proof, starting with the plane hexagon.

Let $T_{12}T_{13}T_{23}T_{21}T_{31}T_{32}$ be a hexagon in a plane $\tau$, with different vertices and different sides, and such that the sides pass alternately through $A_0$ and $B_0$: $B_0 \in T_{12}T_{13}$, $T_{23}T_{21}$, $T_{31}T_{32}$; $A_0 \in T_{13}T_{23}$, $T_{21}T_{31}$, $T_{32}T_{12}$.

Choose $S_{00}$ outside $\tau$; denote $S_{00}A_0$ by $a_0$ and $S_{00}B_0$ by $b_0$.

Choose $S_{01}$ on $a_0$ ($S_{01} \neq S_{00}$, $A_0$) and $S_{11}$ on $S_{00}T_{11}$ ($S_{11} \neq S_{00}$); $S_{01}S_{11} = b_1$. $S_{21} = S_{00}T_{21} \cap b_1$; $S_{31} = S_{00}T_{31} \cap b_1$.

Now take $S_{10}$ on $b_0$ ($S_{10} \neq S_{00}$, $B_0$); $S_{10}S_{11} = a_1$. $S_{12} = S_{00}T_{12} \cap a_1$; $S_{13} = S_{00}T_{13} \cap a_1$.

Next choose $S_{22}$ on $S_{00}T_{22}$, not in the plane through $a_1$ and $b_1$. The plane $S_{22}a_0$ contains $A_0T_{22}$, and thus $T_{12}$ and $T_{32}$. It follows that $S_{22}S_{12}$ intersects $a_0$ and $S_{00}T_{32}$; call these points of intersection $S_{02}$ and $S_{32}$ respectively. In the same manner we find $S_{20}$ and $S_{23}$. $S_{20}S_{22} = a_2$, $S_{02}S_{22} = b_2$. The lines $S_{31}S_{32} = a_3$ and $S_{13}S_{23} = b_3$ are determined. We complete the figure by $S_{03} = a_0 \cap b_3$ and $S_{30} = a_3 \cap b_0$ (these points exist!).

By $R_{16}$, $a_3$ and $b_3$ intersect in $S_{33}$. As $S_{12}$, $S_{13}$, $S_{21}$, $S_{31}$ are in the plane through $a_1$ and $b_1$, $S_{12}S_{21}$ and $S_{13}S_{31}$ intersect. Also $S_{12}S_{21}$ intersects $S_{23}S_{32}$, and $S_{13}S_{31}$ intersects $S_{23}S_{32}$. Since the three lines are not in a plane, they pass through a point $R$. If $R'$ is the projection of $R$ on $\tau$, then $T_{12}T_{21}$, $T_{23}T_{32}$ and $T_{31}T_{13}$ pass through $R'$. This proves the dual of $P_{10}$.

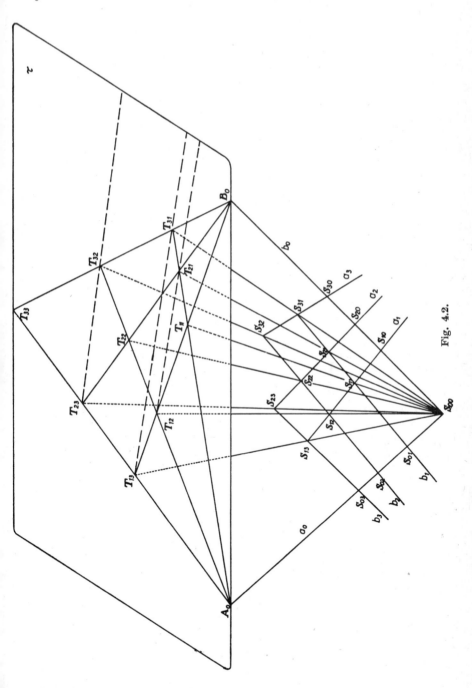

Fig. 4.2.

# CHAPTER V

## COORDINATES IN SPACE

### § 5.1. Coordinates of a point.

In the space $\mathfrak{P}_0^3$ (R1—R5) we choose the points $O$, $X$, $Y$, $Z$, $E$ such that no four of them are in a plane. The planes $XYZ$, $OYZ$, $OZX$, $OXY$ are denoted by $\alpha_0$, $\alpha_1$, $\alpha_2$, $\alpha_3$ respectively. $OE \cap \alpha_0 = W$.

**Definition.** If $P \notin XYZ$, then the coordinates $p$, $q$, $r$ of $P$ in the system $OXYZE$ are

$$p = YZP \cap OE$$
$$q = ZXP \cap OE$$
$$r = XYP \cap OE.$$

The assignment of coordinates to points establishes a one-to-one correspondence between $\mathfrak{P}_0^3 \setminus \alpha_0$ and $(OE \setminus \{W\})^3$.

$x$, $y$, $z$ are variables for the first, second and third coordinates of a point.

The equation of a plane through $YZ$ is $x = c$; analogously for planes through $ZX$ or $XY$.

Let $\tau$ be any plane through $OE$, then $\tau$ is a $\mathfrak{P}^2(D_{11})$. Thus, as in section 3.2, we can form the group $G_a$ of $\Pi_2$'s with one invariant point $W$, and the group $G_m$ of $\Pi_2$'s with invariant points $O$, $W$.

**Definition.** If $A$, $B$, $C$, $D$ are points of $OE \setminus \{W\}$, $B+A = C$ means $(OAW^2)(OBW^2)=(OCW^2)$, and if $A$, $B \neq O$, $B \cdot A = D$ means $(EAO^2W^2)(EBO^2W^2)=(EDO^2W^2)$; $O \cdot A = B \cdot O = O$.

As $D_{11}$ is valid in $\tau$, this is in accordance with the results of section 3.2 (see the proofs of Th. 3.2.1 and Th. 3.2.2).

**Theorem 5.1.1.** If addition and multiplication are defined as in the preceding definition, the points of $OE$ ($W$ excluded) form a division ring.

PROOF. This is an immediate consequence of Th. 3.2.8.

REMARK. Let two planes $\tau$ and $\tau'$ be given, and a point $Z$ not in $\tau$ or $\tau'$. The projection $\pi$ of $\tau$ onto $\tau'$ can be defined by $\pi(A)=ZA \cap \tau'$ for every $A \in \tau$. Then a $\Pi_2$ $(AA'O^2U^2)$ of a line $l$ in $\tau$ onto itself is transformed by $\pi$ into the $\Pi_2((\pi A)(\pi A')(\pi O)^2(\pi U)^2)$ of the line $\pi l$ onto itself, because $\pi$ transforms the lines of $\tau$ into the lines of $\tau'$ and preserves incidence relations.

Now let $\tau$ and $\tau'$ be two planes through $OE$. Projection from $Z$ (or $X$ or $Y$) has the property mentioned in the remark. Thus, $(OAW^2)$ and $(EAO^2W^2)$ are independent of the choice of $\tau$. We conclude that addition and multiplication on $OE$ are independent of $\tau$.

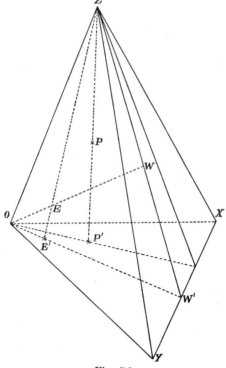

Fig. 5.1.

Moreover, let $P'$ denote the projection of $P$ from $Z$ on $\alpha_3$. If $A$, $B$, $C \in OE$ and $B + A = C$, then $(OAW^2)=(BCW^2)$. By the remark above we see that $(OA'W'^2)=(B'C'W'^2)$, i.e. $B' + A' = C'$

in the coordinate system $OXYE'$ in $\alpha_3$ (section 3.2). In the same way, if $B \cdot A = D$, then $B' \cdot A' = D'$. Thus by projection from $Z$ the division ring of points on $OE$ is mapped isomorphically onto the division ring of points on $OE'$. We shall use this fact by identifying elements of these division rings, which correspond in this mapping. Thus, if $(x, y, z)$ are the coordinates of $P$ in $OXYZE$, then $(x, y)$ are the coordinates of $P'$ in $OXYE'$. Analogous remarks apply to the projections from $X$ on $\alpha_1$ and from $Y$ on $\alpha_2$.

## § 5.2. Equation of a plane.

If $H \in XY$, $H \neq Y$, then in $OXYE'$, $OH$ has an equation of the form $y = xp$. From the preceding section it follows that $y = xp$ is also the equation of the plane $OZH$.

Analogously, the equation of a plane through $OY$ is of the form $z = xq$ and that of a plane through $OX$ is $z = yr$ (which planes

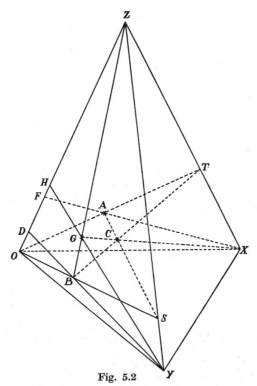

Fig. 5.2

are exceptions?). More generally, let $xa + yb = c$ be the equation
of a line $l$ in $OXY$; then $xa + yb = c$ is also equation of the
plane $Zl$.

The next step is to determine the equation of a plane through
$O$. It is convenient to use here the points of $OZ$, instead of those
of $OE$, as coordinates. For this purpose we project the points
of $OE$ from $XY$ on $OZ$. (i.e. $t \in OE \setminus \{W\}$ is transformed into
$tXY \cap OZ$). In the same way as above, we see that this is an
isomorphic mapping of the division ring $OW$ onto the division
ring $OZ$.

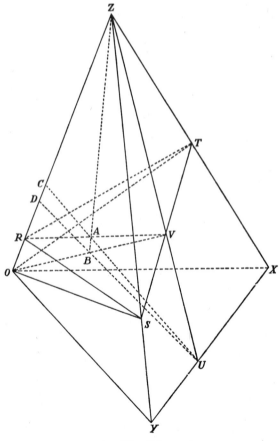

Fig. 5.3.

Let $\alpha$ be a plane through $O$, not through $Z$. $\alpha \cap YZ = S$, $\alpha \cap XZ = T$. Let $z = yq$ and $z = xp$ be the equations of the planes $OXS$ and $OYT$ respectively. Let $C = (x, y, z)$ be any point in $\alpha$, $OT \cap SC = A$, $OS \cap TC = B$. The coordinates of these points are $A = (x, 0, xp)$; $B = (0, y, yq)$. $XC \cap BZ = G$, $YG \cap OZ = H$, $YB \cap OZ = D$, $XA \cap OZ = F$. $D = (0, 0, yq)$; $F = (0, 0, xp)$. It is clear that $G \in FS$. Now project the points of $OZ$ from $S$ on $BZ$, and from $Y$ back on $OZ$; this is the construction of $(ODZ^2)$. We see that $(ODZ^2)F = H$, i.e. $F + D = H$. Thus $H$ has coordinates $(0, 0, xp + yq)$, $G = (0, y, xp + yq)$, and $C = (x, y, xp + yq)$. It follows that the equation of $\alpha$ is $z = xp + yq$.

Now let $\beta$ be a plane not passing through $O$ or $Z$. $\beta \cap OZ = R = (0, 0, r)$, $\beta \cap YZ = S$, $\beta \cap XZ = T$, $OST = \alpha$; let $z = xp + yq$ be the equation of $\alpha$. If $A = (x, y, z)$ is a point of $\beta$, we construct $B = \alpha \cap ZA$. $B = (x, y, xp + yq)$. $OZA \cap XY = U$, $UA \cap OZ = C = (0, 0, z)$. $UB \cap OZ = D = (0, 0, xp + yq)$. Projecting the points of $OZ$ from $B$ on $ZU$ and back from $A$ on $OZ$, we find that $(ORZ^2)D = C$, i.e. $C = D + R$. $z = xp + yq + r$. This is the equation of $\beta$.

Combining the different cases we find

**Theorem 5.2.1.** The equation of a plane ($XYZ$ excluded) has the form $xa + yb + zc = d$, and every equation of this form represents a plane.

## § 5.3. Homogeneous coordinates.

Homogeneous coordinates can be introduced by the same method as in section 3.6. Let $(x, y, z)$ be the coordinates of $P$ and let $x_0$ be an element of the field $\mathfrak{T}$ on $OE$, $x_0 \neq 0$.

Any quadruple $(x_0, x_1, x_2, x_3)$ in the left ratio $(1, x, y, z)$ is a quadruple of homogeneous coordinates for $P$. Also, a quadruple of homogeneous coordinates for the plane $xa + yb + zc = d$ is any quadruple $(\xi_0, \xi_1, \xi_2, \xi_3)$ in the right ratio $(-d, a, b, c)$. The incidence relation between point and plane becomes

$$(1) \qquad x_0\xi_0 + x_1\xi_1 + x_2\xi_2 + x_3\xi_3 = 0.$$

It remains to extend these coordinates to the points of $XYZ$. Let $R$ be a point of $XYZ$, $R \notin XY$, and let $y = zq$, $x = zp$ be the equations of $XOR$, $YOR$ respectively. We give $R$ the homogeneous coordinates $(0, p, q, 1)$. The non-homogeneous coordinates of a point $S$ of $OR$ are $(zp, zq, z)$; its homogeneous coordinates are $(1, zp, zq, z)$, or, equivalently, $(u, p, q, 1)$, where $u = z^{-1}$.

Now let $\alpha$ be a plane with equation (1). $\alpha \cap XYZ = l$; $Ol = \beta$. The equation of $\beta$ is

$$(2) \qquad\qquad x_1\xi_1 + x_2\xi_2 + x_3\xi_3 = 0;$$

this follows from the proof of Th. 5.2.1.

If $R \, \epsilon \, l$ and $S \, \epsilon \, OR$, $R = (0, r_1, r_2, r_3)$, then $S = (s_0, r_1, r_2, r_3)$, as we have just found. As $S$ satisfies (2), $R$ satisfies (2) as well; but then $R$ satisfies (1).

It is left to the reader to draw a similar conclusion for a point $R \, \epsilon \, XY$, $R \neq Y$, to which we give the coordinates $(0, 1, p, 0)$ if $y = xp$ is the plane $OZR$, and for the point $Y = (0, 0, 1, 0)$.

We combine these cases by saying that the points of a plane which lie in $XYZ$, satisfy the equation of that plane. Of course the plane $XYZ$ has the equation $x_0 = 0$.

**Theorem 5.3.1.** In $\mathfrak{P}_0$ (R1—R5), we can assign left-homogeneous coordinates $(x_0, x_1, x_2, x_3)$ to the points and right-homogeneous coordinates $(\xi_0, \xi_1, \xi_2, \xi_3)$ to the planes, such that (1) is the condition for the incidence of point and plane.

Similarly as in § 3.6, the duality principle for the space is a direct consequence of Th. 5.3.1.

Now let $\mathfrak{F}$ be the division ring determined by a projective space $\mathfrak{P}_0^3$ and a coordinate system $OXYZE$ (see Th. 5.1.1). We form the four-dimensional left vector space $\mathfrak{V}_4 = R_4(\mathfrak{F})$ over $\mathfrak{F}$. The representation of the points of $\mathfrak{P}_0^3$ by homogeneous coordinates $(x_0, x_1, x_2, x_3)$ consists in a one-to-one correspondence between the points of $\mathfrak{P}_0^3$ and the one-dimensional subspaces of $\mathfrak{V}_4$. (1) shows that a plane corresponds to a three-dimensional subspace; a line, as the intersection of two planes, corresponds to a two-dimensional subspace.

**Theorem 5.3.2.** Let $\mathfrak{F}$ be the division ring corresponding to the projective space $\mathfrak{P}_0^3$, then $\langle \mathfrak{A}_1, \mathfrak{A}_2, \mathfrak{A}_3, I \rangle$, where $\mathfrak{A}_i$ $(i = 1, 2, 3)$ is the set of $i$-dimensional subspaces of $R_4(\mathfrak{F})$ and $I$ is the incidence relation, is a projective space isomorphic to $\mathfrak{P}_0^3$.

## § 5.4. The geometry over a given division ring.

If $\mathfrak{U}$ and $\mathfrak{W}$ are subspaces in $R_4(\mathfrak{F})$ of different dimensions, then $\mathfrak{U}I\mathfrak{W}$ means that $\mathfrak{U} \subset \mathfrak{W}$ or $\mathfrak{W} \subset \mathfrak{U}$.

It is obvious from the elements of the theory of vector spaces that the projective space $\mathfrak{P}^3_{\mathfrak{F}} = \langle \mathfrak{A}_1, \mathfrak{A}_2, \mathfrak{A}_3, I \rangle$ is a model for the axioms R1,..., R5. $\mathfrak{P}^3_{\mathfrak{F}}$ is called a projective space over $\mathfrak{F}$.

**Exercise.** The reader may verify the axioms for this model.

Now consider a projective plane $\mathfrak{P}^2_0$ (V1, V2, V3, $D_{11}$). Construct the ternary field $\mathfrak{T}$ corresponding to it (section 3.1); by Th. 3.2.9, $\mathfrak{T}$ is a division ring. Next we construct the projective space $\mathfrak{P}^3_{\mathfrak{T}}$. By Th. 4.1.7, the plane $x_3 = 0$ is a projective plane $\mathfrak{P}^2_1$; by the method of section 3.4, we see that $\mathfrak{P}^2_1$ is isomorphic to the given plane $\mathfrak{P}^2_0$. Thus every $\mathfrak{P}^2_0(D_{11})$ can be embedded in a $\mathfrak{P}^3_0$. The converse follows from Th. 4.1.7.

So we have

**Theorem 5.4.1.** A projective plane can be isomorphically embedded in a projective space if and only if $D_{11}$ is valid in it.

If a division ring $\mathfrak{F}$ is given, we can form $\mathfrak{P}^3_{\mathfrak{F}}$ as described above. From Th. 4.1.7 and the end of section 5.1, we infer that every plane of $\mathfrak{P}^3_{\mathfrak{F}}$ will be a $\mathfrak{P}^2_{\mathfrak{F}}$ in which $D_{11}$ is valid. This provides a new proof of Th. 3.4.3.

# CHAPTER VI

# THE FUNDAMENTAL PROPOSITION OF PROJECTIVE GEOMETRY

## § 6.1. The fundamental proposition.

We investigate further the projective plane $\mathfrak{P}^2_\mathfrak{F}$ over a division ring $\mathfrak{F}$. $\mathfrak{P}^2_\mathfrak{F}$ is isomorphic to the geometry of linear subspaces in a three-dimensional left vector space $R_3(\mathfrak{F})$ over $\mathfrak{F}$ (Th. 5.3.2).

A basis for $R_3(\mathfrak{F})$ consists of three independent vectors $\mathbf{a}$, $\mathbf{b}$, $\mathbf{c}$, which can be determined by their lines $a$, $b$, $c$ and the vector $\mathbf{e} = \mathbf{a} + \mathbf{b} + \mathbf{c}$. With respect to the basis, a vector $\mathbf{p} = p_0\mathbf{a} + p_1\mathbf{b} + p_2\mathbf{c}$ has coordinates $(p_0, p_1, p_2)$. The line $p$ is determined by the left ratio $(p_0, p_1, p_2)$, or, what comes to the same, by the quotients

$$p_1^* = p_0^{-1}p_1, \quad p_2^* = p_0^{-1}p_2$$

if we suppose that $p$ is not in the plane through $b$ and $c$.

Let $e$ be the line determined by $\mathbf{e}$. We call $p_1^*$, $p_2^*$ the analytic coordinates of $p$, and also of the point $P$ of $\mathfrak{P}^2_\mathfrak{F}$ which corresponds to $p$. The equation of the plane through $a$ and $e$ is $p_1^* = p_2^*$.

To $a$, $b$, $c$, $e$ there correspond points $O$, $X$, $Y$, $E$. A point on $OE$ has analytic coordinates $(p_1^*, p_1^*)$. In the geometric coordinate system $OXYE$, $P$ has coordinates $p_1^0$, $p_2^0$, where $p_1^0$, $p_2^0$ are points on $OE$, and $p_1^0 = (p_1^*, p_1^*)$, $p_2^0 = (p_2^*, p_2^*)$. The points on $OE$ form the division ring $\mathfrak{F}_1$ attached to the geometry $\mathfrak{P}^2_\mathfrak{F}$ by the coordinate system $OXYE$ (see section 3.1). By the same method as in section 3.4 we find that the correspondence $\sigma(x) = (x, x)$ is an isomorphism between $\mathfrak{F}$ and $\mathfrak{F}_1$. We shall regard this isomorphism as the identity and thus write "$x$" for "$(x, x)$".

Thus the geometric coordinates of $P$ in $OXYE$ are $(p_1^*, p_2^*)$, or, what is the same, $(p_0^{-1}p_1, p_0^{-1}p_2)$. Introducing homogeneous coordinates, we find

**Theorem 6.1.1.** Let $P$ be the point of $\mathfrak{P}^2_\mathfrak{F}$, corresponding to the line $p$ in $R_3(\mathfrak{F})$. Let $\mathbf{a}$, $\mathbf{b}$, $\mathbf{c}$ be a basis of $R_3(\mathfrak{F})$ and $\mathbf{e}$ the vector $\mathbf{a} + \mathbf{b} + \mathbf{c}$. Let the points $O$, $X$, $Y$, $E$ correspond to the lines $a$, $b$, $c$, $e$ of $R_3(\mathfrak{F})$ which are determined by $\mathbf{a}$, $\mathbf{b}$, $\mathbf{c}$, $\mathbf{e}$. Then the analytic coordinates of $p$ with respect to the basis $\mathbf{a}$, $\mathbf{b}$, $\mathbf{c}$ are the same as the geometric coordinates of $P$ in the system $OXYE$.

Now let $\mathbf{a}'$, $\mathbf{b}'$, $\mathbf{c}'$ be a different basis of $R_3(\mathfrak{F})$, and $(p'_0, p'_1, p'_2)$ the coordinates of $p$ with respect to this basis. The relation between the two kinds of coordinates is given by the following coordinate transformation in $R_3(\mathfrak{F})$ (section 1.6, (4)):

$$(1) \qquad\qquad x_k = \sum_{i=0}^{2} x'_1 u^i_k.$$

$\mathbf{a}'$, $\mathbf{b}'$, $\mathbf{c}'$ determine a geometric coordinate system $O'X'Y'E'$; the points of $O'E'$ form a division ring $\mathfrak{F}_2$. Again there is an isomorphism between $\mathfrak{F}$ and $\mathfrak{F}_2$, which will be regarded as the identity. By this method we obtain a privileged isomorphism between $\mathfrak{F}_1$ and $\mathfrak{F}_2$, which we again regard as the identity, so that every element of $\mathfrak{F}_1$ is identified with one and only one element of $\mathfrak{F}_2$.

Now let $l$ and $l'$ be lines in $\mathfrak{P}^2_\mathfrak{F}$ and let a perspectivity $\mathfrak{S}$ be given between $l$ and $l'$. $\mathfrak{S}$ can be extended to a central collineation $\mathfrak{C}$ (Th. 2.5.6). $\mathfrak{C}$ transforms the coordinate system $OXYE$ into another system which we call $O'X'Y'E'$. As above, let $\mathfrak{F}_1$ and $\mathfrak{F}_2$ be the division rings attached to $\mathfrak{P}^2_\mathfrak{F}$ by these systems. As was proved in Th. 3.3.1, $\mathfrak{C}$ establishes an isomorphism $\sigma$ between $\mathfrak{F}_1$ and $\mathfrak{F}_2$, such that, if $(p_0, p_1, p_2)$ are the coordinates of $P$ in $OXYE$, then $(\sigma p_0, \sigma p_1, \sigma p_2)$ are the coordinates of $\mathfrak{C}P$ in $O'X'Y'E'$. Please take notice that $\sigma$ need not be the isomorphism which we have regarded as the identity! (See the end of this section.)

Let us denote $\mathfrak{C}P$ by $P^*$ and the coordinates of $P^*$ in $OXYE$ by $p^*_i$; those in $O'X'Y'E'$ by $p^{*\prime}_i$. Then $p^{*\prime}_i = \sigma p_i$. Now (1) gives us

$$(2) \qquad\qquad p^*_k = \sum_{i=0}^{2} (\sigma p_i) u^i_k.$$

This is the analytical expression of $\mathfrak{C}$ in $OXYE$.

As $\mathfrak{F}$ has been identified with $\mathfrak{F}_1$ as well as with $\mathfrak{F}_2$, $\sigma$ can be considered as an automorphism of $\mathfrak{F}$. We shall prove that $\sigma$ is an inner automorphism of $\mathfrak{F}$. (This proof is due to R. Baer, see: Linear Algebra and Projective Geometry (New York 1952), section III 1.)

First of all, we draw some simple conclusions from (2).

$$(x_k + y_k)^* = \sum_{i=k}^{2} \sigma(x_i + y_i)u_k^i = \sum (\sigma x_i)u_k^i + \sum (\sigma y_i)u_k^i = x_k^* + y_k^*. \text{ So}$$

$$(3) \qquad\qquad (x_k + y_k)^* = x_k^* + y_k^*.$$

$$(\lambda x_k)^* = \sum \sigma(\lambda x_i)u_k^i = \sum(\sigma\lambda)(\sigma x_i)u_k^i = (\sigma\lambda)\sum(\sigma x_i)u_k^i = (\sigma\lambda)x_k^*. \text{ So}$$

$$(4) \qquad\qquad (\lambda x_k)^* = (\sigma\lambda)x_k^*.$$

Now let $P$ be a point on the axis $d$ of $\mathfrak{C}$, thus $P^* = P$ and $p_i^*$ is proportional to $p_i : p_i^* = \mu(P)p_i$. We shall show that $\mu(P)$ does not depend upon $P$.

If $Q$ is another point on $d$, then $(p_i + q_i)$ are also the coordinates of a point on $d$, denoted by $P + Q$.

$$(5) \qquad\qquad (p_i + q_i)^* = \mu(P + Q)(p_i + q_i).$$

Also, according to (3) and (4):

$$(6) \qquad (p_i + q_i)^* = p_i^* + q_i^* = \mu(P)p_i + \mu(Q)q_i.$$

By (5) and (6):

$$\{\mu(P + Q) - \mu(P)\}p_i + \{\mu(P+Q) - \mu(Q)\}q_i = 0 \quad (i = 0, 1, 2).$$

Because $P$ and $Q$ are different points, this implies $\mu(P) = \mu(Q) = \mu(P + Q)$. We now write simply $\mu$ instead of $\mu(P)$.

For any point $P$ on the axis we have $(\lambda p_k)^* = \mu(\lambda p_k)$, and by (4) : $(\lambda p_k)^* = (\sigma\lambda)p_k^* = (\sigma\lambda)\mu p_k$. Thus $\mu\lambda p_k = (\sigma\lambda)\mu p_k$; $\mu\lambda = (\sigma\lambda)\mu$; $\sigma\lambda = \mu\lambda\mu^{-1}$. As this is true for every $\lambda$ in $\mathfrak{F}$, $\sigma$ is an inner automorphism of $\mathfrak{F}$.

We can now illustrate the fact that, if $\mathfrak{F}$ is not commutative, $\sigma$ need not be the identity. Let $a$ be an element of $\mathfrak{F}$ such that $ax = xa$ is not true for every $x$ in $\mathfrak{F}$. The automorphism $x^* = axa^{-1}$ defines a projectivity $\sigma$ on $OE$ (see Exercise 1 at the end of the section); this projectivity can be extended to a collineation $\mathfrak{C}$ (Cor. 2.5.6). $\mathfrak{C}O = O$, $\mathfrak{C}E = E$, $\mathfrak{C}W = W$, $\mathfrak{C}X = X'$, $\mathfrak{C}Y = Y'$. Coordinates in $OX'Y'E$ are the same as in $OXYE$ (see the proof of Th. 3.3.1), but $\sigma$ is not the identity.

$\sigma$ is a projectivity on $OE$ with three invariant points, which is not the identity. Conversely, the result of the preceding pages shows that every projectivity on $OE$ which leaves $O$, $E$, $W$ invariant, induces an inner automorphism of $\mathfrak{F}$. Now $\mathfrak{F}$ has inner automorphisms different from the identity if and only if $\mathfrak{F}$ is not commutative. Thus in $\mathfrak{P}^2_{\mathfrak{F}}$ there exist non-identical projectivities with three invariant points if and only if $\mathfrak{F}$ is not commutative. In other words

**Theorem 6.1.2.** If $\mathfrak{F}$ is commutative, then every projectivity on a line in $\mathfrak{P}^2_{\mathfrak{F}}$ with three invariant points is the identity. Conversely: If in $\mathfrak{P}^2_{\mathfrak{F}}$ every projectivity on a line with three invariant points is the identity, then $\mathfrak{F}$ is commutative.

The proposition that every projectivity on a line with three invariant points is the identity, is equivalent to the proposition that a projectivity between two lines is uniquely determined by the images of three points. For if $\sigma$ and $\tau$ are projectivities from $l$ onto $l'$ such that $\sigma A = \tau A$, $\sigma B = \tau B$, $\sigma C = \tau C$, then $\tau^{-1}\sigma$ is a projectivity on $l$ which leaves $A$, $B$, $C$ invariant, and $\tau^{-1}\sigma$ is the identity if and only if $\sigma = \tau$.

The proposition that a projectivity between two lines is uniquely determined by the images of three points, is called the fundamental proposition of projective geometry. (It got this name because it played a fundamental part in the synthetic construction of projective geometry.) (See Appendix 4.)

**Theorem 6.1.3.** In $\mathfrak{P}(D_{11})$, the fundamental proposition of projective geometry is equivalent to the commutative law of multiplication in $\mathfrak{T}(\mathfrak{P})$, and thus also the Pappos' proposition.

**Exercise 1.** The projectivity $x^* = a^{-1}xa$ is constructed as follows: $EY \cap aX = C$, $OC \cap EX = D$, $xY \cap OC = Q$, $XQ \cap EY = R$, $OR \cap DY = S$, $XS \cap OE = a^{-1}xa$. This comes to the following construction: Project $x$ from $Y$ on $OC$, from $X$ on $EY$, from $O$ on $DY$, from $X$ on $OE$. This is a $\Pi_4$. (See fig. 6.1, which is drawn inaccurately, since otherwise $x^*$ would coincide with $x$). The last statement also follows from the fact that multiplication to the right is effectuated by a $\Pi_2$, multiplication to the left by a $\Pi_3$ (see the proofs of Th. 3.2.6 and 3.2.7). Two of the projections here can be coupled into one.

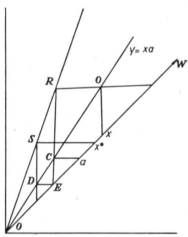

Fig. 6.1.

**Exercise 2.** If $\mathfrak{F}$ is the division ring of real quaternions $c_0e + c_1j_1 + c_2j_2 + c_3j_3$, and $x^* = j_1^{-1}xj_1$, then every point of the form $c_0e + c_1j_1$ is invariant, but $j_1^{-1}j_2j_1 = -j_1j_2j_1 = -j_2$; $j_1^{-1}j_3j_1 = -j_1j_3j_1 = -j_3$.

## § 6.2. Summary of results.

The following axiom systems for plane geometry have been considered.

$$S_1 = \{V1, V2, V3\}.$$
$$S_2 = \{V1, V2, V3, D_{10}\}.$$
$$S_3 = \{V1, V2, V3, D_{11}\}.$$
$$S_4 = \{V1, V2, V3, P_{10}\}.$$

All these systems are consistent, because analytic plane projective geometry $APPG$ is a model for them.

(1)                              $S_1 \subset S_2 \subset S_3 \subset S_4.$

This was proved in Th. 2.2.13, Th. 3.5.1, Th. 3.5.2 respectively.

We shall not prove the independence of the axioms in $S_1$. Isolated independence results are contained in (1).

$S_4$ is not complete with respect to $APPG$. To prove this, we recall that the projective plane over an arbitrary field $\mathfrak{F}$ is a model of $S_4$ (Th. 3.4.4). Let us choose for $\mathfrak{F}$ the prime field $\mathfrak{F}_2$ of characteristic 2, that is the field consisting of two elements 0, 1 with as definitions of $+$ and $\times$:

$$0 + 0 = 0, \ 0 + 1 = 1, \ 1 + 0 = 1, \ 1 + 1 = 0,$$
$$0 \times 0 = 0, \ 0 \times 1 = 0, \ 1 \times 0 = 0, \ 1 \times 1 = 1.$$

The projective plane over $\mathfrak{F}_2$ contains seven points: $A_1 = (1, 0, 0)$, $A_2 = (0, 1, 0)$, $A_3 = (0, 0, 1)$, $A_4 = (0, 1, 1)$, $A_5 = (1, 0, 1)$, $A_6 = (1, 1, 0)$, $A_7 = (1, 1, 1)$. There are also seven lines: $l_1 = A_2 A_3 A_4$, $l_2 = A_1 A_3 A_5$, $l_3 = A_1 A_2 A_6$, $l_4 = A_1 A_4 A_7$, $l_5 = A_2 A_5 A_7$, $l_6 = A_3 A_6 A_7$, $l_7 = A_4 A_5 A_6$.

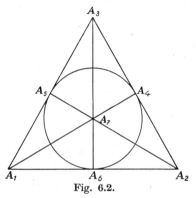

Fig. 6.2.

This geometry is visualized in fig. 6.2, where the circle represents $l_7$. The harmonic conjugate of $A_5$ with respect to $A_1 A_3$ can be constructed by means of the quadrangle $A_2 A_4 A_6 A_7$; it is $A_5$ itself.

More generally, in a projective plane $\mathfrak{P}_0(D_{11})$, with a coordinate system $OXYE$, let us construct the harmonic conjugate $F$ of $E$ with respect to $O$, $W$. If $XE \cap OY = P$, $WP \cap OX = Q$, then $YQ \cap OE$ is $F$, for in the quadrangle $XYPQ$ we have $EF$ harm. $OW$. The figure shows that $E + F = O$, so $F = E$ if and only if the characteristic of $\mathfrak{T}(\mathfrak{P}_0)$ is 2.

Therefore the following theorem of $APPG$ is not provable in $S_4$:

**Harmonic proposition** $(H)$. If $A$, $B$, $C$ are different points on a line $l$, and if $AB$, $CD$ are harmonic pairs, then $D \neq C$.

The projective plane over any field is a model of $S_4$. If two fields are not isomorphic, the corresponding projective planes are not isomorphic (this follows from Th. 3.3.1). As there is an infinity of non-isomorphic fields, it follows that $S_4$ is not categorical.

# CHAPTER VII

## ORDER

### § 7.1. Cyclically ordered sets.

As we saw in § 1.4, the points of a line in analytic projective geometry can be cyclically ordered, in such a way that this order is invariant under projection. The cyclical order could be described by means of a relation between pairs of points, called separation.

We shall now introduce the notion of cyclical order in axiomatic geometry; it is convenient to treat first the general theory of cyclically ordered sets, independently of its application to geometry.

**Definition.** A cyclically ordered set is a set $V$ together with a relation $\sigma$ between pairs of elements of $V$, such that the following axioms C1—C4 are satisfied.

**Axioms.**

**C1.** If $AB\sigma CD$, then $A$, $B$, $C$ and $D$ are distinct elements of $V$.

**C2.** $AB\sigma CD$ is equivalent to $CD\sigma AB$ and to $BA\sigma CD$. In other words, $\sigma$ is a symmetric relation between unordered pairs of elements of $V$.

**C3.** If $A$, $B$, $C$, $D$ are distinct elements of $V$, then one and only one of the relations $AB\sigma CD$, $AC\sigma BD$, $AD\sigma BC$ holds.

If $AB\sigma CD$, then $AB$ and $CD$ are called *separating* pairs. Thus four distinct elements of $V$ can be divided in one and only one way into separating pairs.

**C4.** If $AC\sigma BD$ and $AD\sigma CE$, then $AD\sigma BE$.

**Theorem 7.1.1.** In $V$(C1—C4): If $A$, $B$, $X$, $Y$, $Z$ are distinct points, and $AB\sigma XZ$, but not $AB\sigma YZ$, then $AB\sigma XY$.

PROOF. By C3 we have either $AB\ \sigma\ XY$ or $AX\ \sigma\ BY$ or $AY\ \sigma\ BX$. First suppose $AX\ \sigma\ BY$. From $AX\ \sigma\ YB$ and $AB\ \sigma\ XZ$ we infer, by C4, $AB\ \sigma\ YZ$, contrary to the hypothesis. Next suppose $AY \, \sigma \, BX$. From $BX \, \sigma \, YA$ and $BA \, \sigma \, XZ$ we infer, similarly, $BA \, \sigma \, YZ$, which is not so. Thus $AX \, \sigma \, BY$ and $AY \, \sigma \, BX$ are both false, and by C3 we have $AB \, \sigma \, XY$.

In order to give a formulation of Th. 7.1.1 whith is more convenient to apply, we introduce the relation $\rho$ as follows:

**Definition.** If $A$, $B$, $C$, $D \in V$, then $AB \, \rho \, CD$ means that $A$ and $B$ are distinct from $C$ and $D$, and that $AB \, \sigma \, CD$ is false. (Thus it is allowed that $A = B$ or (and) $C = D$.)

It is easy to deduce the following theorem from Th. 7.1.1:

**Theorem 7.1.2.** In $V(\text{C1}-\text{C4})$: If $AB \, \rho \, XY$ and $AB \, \sigma \, XZ$, then $AB \, \sigma \, YZ$.

While C4 is ill-adapted for use, Th. 7.1.2 is easy to visualize and to apply, the main reason for this difference being that in this theorem one of the pairs, namely $AB$, is fixed. Th. 7.1.2 can fully replace C4, for C4 can be derived from C1, C2, C3 and Th. 7.1.2. (**Exercise**: this proof is left to the reader). Therefore we shall rarely use C4 in proofs; in most cases the use of Th. 7.1.2 is more convenient.

There are two theorems parallel to Th. 7.1.2.

**Theorem 7.1.3.** In $V(\text{C1}-\text{C4})$: If $AB \, \rho \, XY$ and $AB \, \rho \, XZ$, then $AB \, \rho \, YZ$.

PROOF. $AB \, \sigma \, YZ$ leads immediately to a contradiction, by Th. 7.1.2.

**Theorem 7.1.4.** In $V(\text{C1}-\text{C4})$: If $AB \, \sigma \, XY$ and $AB \, \sigma \, XZ$, then $AB \, \rho \, YZ$.

PROOF. Suppose $AB \, \sigma \, YZ$; then $A$, $X$, $Y$, $Z$ are all distinct. If, in addition, $AX \, \sigma \, YZ$, then this together with $AX \, \rho \, BY$ would give $AX \, \sigma \, BZ$, in contradiction to $AB \, \sigma \, XZ$. By interchanging $X$, $Y$, $Z$ in this argument, we find that $AY \, \sigma \, XZ$ and $AZ \, \sigma \, XY$ are false as well, which contradicts C3. As the contradiction was derived from the hypothesis $AB \, \sigma \, YZ$, this relation is false, and hence, as is easy to see, $AB \, \rho \, YZ$.

We now choose fixed distinct elements $A$, $B$ and consider the relation between $X$ and $Y$ which is given by $AB \, \rho \, XY$. Let us denote this relation, which is defined on $V \setminus \{A, B\}$, by $\delta_{AB}$.

**Theorem 7.1.5.** In $V(C1-C4)$: The relation $\delta_{AB}$ is reflexive, symmetric and transitive.

PROOF. Reflexivity and symmetry follow immediately from the definition of the relation $\rho$; transitivity is the content of Th. 7.1.3.

COROLLARY 7.1.5. $V \setminus \{A, B\}$ can be divided into equivalence classes with respect to $\delta_{AB}$.

**Theorem 7.1.6.** In $V(C1-C4)$: There are in $V \setminus \{A, B\}$ at most two equivalence classes with respect to $\delta_{AB}$.

PROOF. $X$ and $Y$ belong to different equivalence classes if and only if $AB \, \sigma \, XY$; thus the theorem follows immediately from Th. 7.1.4.

**Definition.** Each of the equivalence classes with respect to $\delta_{AB}$ is called a *segment* $AB$.

Let $\Sigma$ be one of the segments $AB$. Let us now consider the relation between elements of $\Sigma$, which is given by $AY \, \sigma \, BX$, and denote it by $<_{AB}$.

**Theorem 7.1.7.** In $V(C1-C4)$: $<_{AB}$ is an order relation on $\Sigma$.

PROOF. The following has to be proved:

(i)    If $X, Y \in \Sigma$, then one and only one of the relations $X = Y$, $X <_{AB} Y$, $Y <_{AB} X$ holds.

(ii)   If $X <_{AB} Y$ and $Y <_{AB} Z$, then $X <_{AB} Z$.

For elements of $\Sigma$, $XY \, \rho \, AB$, so (i) is an immediate consequence of Axiom C3.

As to (ii), it is given that $AY \, \sigma \, BX$, $AZ \, \sigma \, BY$. The latter relation gives $AY \, \rho \, BZ$, so, by Th. 7.1.2, $AY \, \sigma \, XZ$. This implies $X \neq Z$ and $AZ \, \rho \, XY$. This, with $AZ \, \sigma \, BY$, gives $AZ \, \sigma \, BX$, which is $X <_{AB} Z$.

REMARK. $\delta_{BA}$ is the same relation as $\delta_{AB}$, but $<_{BA}$ is the *opposite* of $<_{AB}$: If $X <_{AB} Y$, then $Y <_{BA} X$, and conversely. We shall say that $<_{AB}$ orders $\Sigma$ *from $A$ to $B$*; $<_{BA}$ orders $\Sigma$ *from $B$ to $A$*.

Now let $\Sigma_1, \Sigma_2$ be the two segments $AB$ (one or even both of them may be empty). Let $\Sigma_{iAB}$ denote $\Sigma_i$, ordered from $A$ to $B$. We form the ordered sum $\{A\} + \Sigma_{1AB} + \{B\} + \Sigma_{2BA}$ (see

§ 1.2); that is, we order $V$ by taking $A$ as the first element, followed by $\Sigma_1$ ordered from $A$ to $B$, then $B$, then $\Sigma_2$ ordered from $B$ to $A$. Let $\gamma'_{AB}$ denote the set $V$, ordered in this way, and $\gamma''_{AB}$ the ordered set $\{A\} + \Sigma_{2AB} + \{B\} + \Sigma_{1BA}$. Evidently, $\gamma'_{AB}$ and $\gamma''_{AB}$ define opposite orders on $V \setminus \{A\}$ but in both $A$ is taken as the first element.

In order to visualize the relation between $\sigma$, $\gamma'_{AB}$ and $\gamma''_{AB}$, we represent $V$, cyclically ordered by $\sigma$, by a circle; if this circle is cut open at $A$, the cyclical order leads to the simple orders $\gamma'_{AB}$ and $\gamma''_{AB}$.

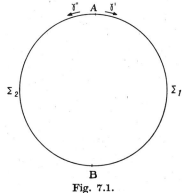

Fig. 7.1.

While in the preceding pages we derived a simple order from a cyclical order, we shall see presently that it is also possible to derive a cyclical order from a simple order.

**Definition.** If $<$ is an order relation in a set $V$, and $P, Q, R, S$ are elements of $V$ such that

(1) $$P < Q < R < S,$$

then $PR$ and $QS$ are *alternating pairs* with respect to $<$; we write $PR \, \tau \, QS$.

$\tau$ is considered as a symmetric relation between the unordered pairs $PR$ and $QS$; that is, $PR \, \tau \, QS$ is true not only if (1) holds, but also if one of the sets of order relations holds which result from (1) by interchanging $P$ and $R$ or $Q$ and $S$ or by interchanging the pairs $PR$, $QS$, or by a combination of these operations. In other words, $PR \, \tau \, QS$ if and only if exactly one of the points $Q, S$ is between $P$ and $R$. Thus four distinct elements of $V$ can be divided into alternating pairs in just one way.

**Theorem 7.1.8.** The relation $\tau$ between alternating pairs in an ordered set $V$ determines a cyclical order of $V$, that is, $\tau$ satisfies axioms C1—C4.

Proof. For C1, C2, C3 this is clear.

Instead of C4 we prove the property contained in Th. 7.1.2. Thus let $AB \, \tau \, XY$, while $AB \, \tau \, YZ$ is false, and $A$, $B \neq Z$. We wish to prove that $AB \, \tau \, XZ$. Just one of $X$, $Y$ is between $A$ and $B$. If $X$ is between A and B, then $Y$ is not, and $Z$ is not, because $AB \, \tau \, YZ$ is false. If $Y$ is between $A$ and $B$, then so is $Z$. In both cases, just one of $X$ and $Z$ is between $A$ and $B$.

In a cyclically ordered set $V$ with cyclical order relation $\sigma$ we choose distinct elements $A$, $B$ and define a simply ordered sum $\gamma'_{AB}$ as above. In $\gamma'_{AB}$, the relation between alternating pairs forms a cyclical order relation $\tau$. It is intuitively clear that $\tau = \sigma$. This is expressed in the following theorem:

**Theorem 7.1.9.** Let the set $V$ be cyclically ordered by the relation $\sigma$; let $A$ and $B$ be distinct elements of $V$. Then alternating pairs in $\gamma'_{AB}$ (or $\gamma''_{AB}$) are separating pairs in $\sigma$.

Proof. It suffices to prove that $P < Q < R < S$ in $\gamma'_{AB}$ entails $PR \, \sigma \, QS$.

According to the definition of $\gamma'_{AB}$, several cases must be distinguished.

I.    $P, Q, R, S \in \Sigma_1$.

II.    $P, Q, R \in \Sigma_1$, $S \in \Sigma_2$.

III.    $P, Q \in \Sigma_1$, $R, S \in \Sigma_2$.

IV.    $P \in \Sigma_1$, $Q, R, S \in \Sigma_2$.

V.    $P, Q, R, S \in \Sigma_2$,

and, in addition, those cases in which one of the elements $P$, $Q$, $R$, $S$ coincides with $A$ or with $B$.

**Case I.** We have (a) $AQ \, \sigma \, BP$; (b) $AR \, \sigma \, BP$; (c) $AS \, \sigma \, BP$; (d) $AR \, \sigma \, BQ$; (e) $AS \, \sigma \, BQ$; (f) $AS \, \sigma \, BR$.

Bij (a), $BQ \, \rho \, AP$; by (d), $BQ \, \sigma \, AR$; thus Th. 7.1.2 gives (g) $BQ \, \sigma \, PR$.

By (f), $BS \, \rho \, AR$; by (c), $BS \, \rho \, AP$; thus Th. 7.1.3 gives (h) $BS \, \rho \, PR$.

By (g), $PR \, \sigma \, BQ$; by (h), $PR \, \rho \, BS$; thus Th. 7.1.2 gives $PR \, \sigma \, QS$.

**Case II.** We have (a) $AQ \, \sigma \, BP$; (b) $AR \, \sigma \, BP$; (c) $AB \, \sigma \, PS$; (d) $AR \, \sigma \, BQ$; (e) $AB \, \sigma \, QS$; (f) $AB \, \sigma \, RS$.

The reader is requested to verify that the proof of case I can be literally repeated here.

**Case III.** We have (a) $AQ \, \sigma \, BP$; (b) $AB \, \sigma \, PR$; (c) $AB \, \sigma \, PS$; (d) $AB \, \sigma \, QR$; (e) $AB \, \sigma \, QS$; (f) $AR \, \sigma \, BS$.

The reader may derive the proof from that in case I by using the same pairs of elements, only interchanging the letters $\rho$ and $\sigma$ in some places.

**Case IV.** This case is reduced to case II by considering $\gamma''_{AB}$ instead of $\gamma'_{AB}$.

**Case V.** This case is reduced to case I in the same way. The more special cases are left to the reader.

COROLLARY 7.1.9: In a set $V$ with cyclical order relation $\sigma$, the simple orderings $\gamma'_{AB}$ and $\gamma'_{CD}$ determine the same relation between alternating pairs. In particular, this is true for $\gamma'_{AB}$ and $\gamma'_{AC}$, where $C$ is any point distinct from $A$.

**Theorem 7.1.10.** Let $\gamma'$ and $\gamma''$ denote the same set $V$, simply ordered by the relations $<'$ and $<''$ respectively. If $\gamma'$ and $\gamma''$ have the same first element $A$ and determine the same relation $\tau$ between alternating pairs, then either $\gamma' = \gamma''$, or $<'$ and $<''$ are opposite orders in $V \setminus \{A\}$.

PROOF. Let $C$ and $D$ be elements of $V$, distinct from $A$, such that $C <' D$. Let $X$ be any element of $V$, distinct from $A, C, D$. One of the following cases necessarily occurs:

(1) $\qquad AX \, \tau \, CD$ or $AC \, \tau \, DX$ or $AD \, \tau \, CX$,

which entail

(2) $A <' C <' X <' D$ or $A <' X <' C <' D$ or $A <' C <' D <' X$.

Thus it is determined by $\tau$ whether $C <' X$ or $X <' C$. If $Y$ is a new element of $V$, it is seen by the same reasoning that $\tau$ determines whether $X <' Y$ or $Y <' X$.

As to $\gamma''$, we distinguish between two cases. If $C <'' D$, then it follows from the preceding reasoning that $X <' Y$ entails $X <'' Y$ and conversely, so that $\gamma' = \gamma''$. If $D <'' C$, instead of (2) we get

(3) $A <'' D <'' X <'' C, \quad A <'' D <'' C <'' X, \quad A <'' X <'' D <'' C.$

Comparing (2) with (3) we see that $X <' C$ if $C <'' X$ and conversely. Repeating this argument, we see that $X <' Y$ if $Y <'' X$ and conversely. Thus $<''$ is opposite to $<'$. This proves the theorem.

COROLLARY 7.1.10. In a set $V$ with cyclical ordering $\sigma$ there are exactly two simple orders $\gamma'$ and $\gamma''$ with first element $A$ and such that the relation between alternating pairs of $\gamma'$ or $\gamma''$ coincides with $\sigma$. In other words, $\gamma'_{AB}$ and $\gamma''_{AB}$ are independent of the choice of $B$.

This corollary follows immediately from the preceding theorem and the corollary to Th. 7.1.9.

## § 7.2. Cyclical order of the projective line.

In order to apply the theory of cyclical order in geometry, we shall assume that every line is cyclically ordered; for this assumption not to be trivial, every line must contain at least four distinct points. Moreover, cyclical order must be invariant under projection.

**Definition.** An ordered plane projective geometry is a quadruple $\langle \Pi, \Lambda, I, \sigma \rangle$, where $\Pi$ and $\Lambda$ are disjoint sets, $I$ is a relation as described in section 2.1, and $\sigma$ is a relation between collinear pairs of points, such that axioms V1, V2, V3, C1—C6 are satisfied.

**Axioms C1—C4.**

Axioms C1—C4 of section 7.1 are valid for every line considered as the set $V$.

C5. At least one line is incident with at least four distinct points.

C6. If $A$, $B$, $C$, $D$ are different points of a line $l$, and $A'$. $B'$, $C'$, $D'$ their projections on $l'$ from some centre $S$, then $AC \, \sigma \, BD$ entails $A'C' \, \sigma \, B'D'$.

**Definition.** If $a$, $b$, $c$, $d$ are lines through $S$, which $l$ intersects in $A$, $B$, $C$, $D$, such that $AC \, \sigma \, BD$, then we write $ac \, \sigma \, bd$.

Note that this definition is justified by axiom C6.

As an immediate corollary to C6 we have:

**Theorem 7.2.1.** In $\mathfrak{P}(C1-C6)$: If $A$, $B$, $C$, $D$ are points on $l$, such that $AC \, \sigma \, BD$, and if $A'$, $B'$, $C'$, $D'$ are their images on $l'$ by a projectivity $\varphi$, then $A'C' \, \sigma \, B'D'$. In other words: cyclical order is invariant under projectivities.

**Theorem 7.2.2.** In $\mathfrak{P}(C1-C6)$: If $A$, $B$, $C$ are distinct points on a line $l$, then $l$ contains a point $D$ such that $AC \, \sigma \, BD$.

Proof. By C5 there is a line $l'$ which contains four distinct points; by C3 these can be arranged in separating pairs $PR$, $QS$. It is easy to construct a projectivity $\varphi$ such that $\varphi P = A$, $\varphi Q = B$, $\varphi R = C$. If $\varphi S = D$, then $AC \, \sigma \, BD$.

Corollary 7.2.2. 1. Every line contains at least four distinct points.

Corollary 7.2.2. 2. If $A$ and $B$ are distinct points on $l$, then there are two non-empty segments $AB$. For, if $C$ is a point on $l$,

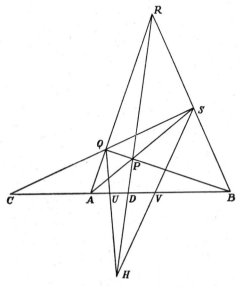

Fig. 7.2.

distinct from $A$ and $B$, and we construct $D$ such that $AB \, \sigma \, CD$, then $C$ and $D$ are on different segments $AB$. By Th. 7.1.6 there are at most two segments $AB$, so the number of segments is exactly two.

EXERCISE. Prove that every segment contains an infinite number of points.

**Theorem 7.2.3.** In $\mathfrak{P}(\text{C1}-\text{C6})$: If $A$, $B$, $C$, $D$ are distinct points on a line $l$, and $PQRS$ is a complete quadrangle such that $A \in PS$, $QR$; $B \in PQ$, $RS$; $C \in QS$; $D \in PR$, then $AB \, \sigma \, CD$.

PROOF. Choose $H$ on $PR$, such that     (1) $PH \, \sigma \, DR$.
Project (1) from $Q$ on $l$:            (2) $BU \, \sigma \, DA$.
Project (1) from $S$ on $l$:             (3) $AV \, \sigma \, DB$.

$AD \, \sigma \, BU$ and $AD \, \rho \, BV$ (from (3)) give $AD \, \sigma \, UV$. By projecting from $H$ on $AS$, and back from $Q$ on $l$, we obtain $AB \, \sigma \, UC$. This together with $AB \, \rho \, UD$ (from (2)) gives $AB \, \sigma \, CD$.

COROLLARY 7.2.3. 1. In $\mathfrak{P}(D_9)$ we have the notion of harmonic pairs, see the remark after Th. 2.4.5. So in $\mathfrak{P}(D_9, \text{C1}-\text{C6})$ harmonic pairs of points are separating pairs of points.

COROLLARY 7.2.3. 2. By Th. 2.4.6, $D_{10}$ is valid in $\mathfrak{P}(D_9, \text{C1}-\text{C6})$.

## § 7.3. Order and coordinates.

The relation between the cyclical order relation in $\mathfrak{P}$ and the properties of $\mathfrak{T}(\mathfrak{P})$ has been investigated by Sybilla Crampe [Math. Zeitschr. 69 (1958), p. 435–462]. (See Appendix 5).

Consider, in $\mathfrak{P}(\text{C1}-\text{C6})$, a coordinate system $OXYE$; $OE \cap XY = W$. Let $\Sigma_1$ be the segment $OW$ which contains $E$, $\Sigma_2$ the other segment $OW$. By $<$ we denote the order relation in

$$\gamma'_{WO} = \{W\} + \Sigma_{2WO} + \{O\} + \Sigma_{1OW}.$$

**Theorem 7.3.1.** In $\mathfrak{P}(\text{C1}-\text{C6})$: If $n < n'$, then

$$\Phi(a, m, n) < \Phi(a, m, n').$$

PROOF. If $a$ and $m$ are fixed, then $\varphi(n) = \Phi(a, m, n)$ is obtained from $n$ by three projections: $\pi_1$ from $X$ on $OY$, $\pi_2$ from $R$ on $aY$, $\pi_3$ from $X$ back on $OW$. Here $R$ is the intersection of the line $y = \Phi(x, m, n)$ with $XY$; $R$ is independent of $n$. See § 3.1, (8) and fig. **7.3**.

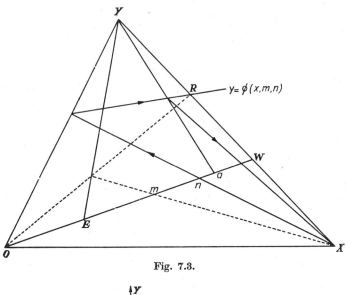

Fig. 7.3.

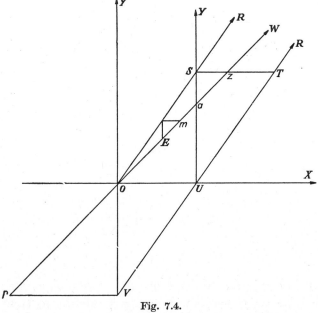

Fig. 7.4.

The cyclical order is invariant under these projections. Moreover, $W$ is invariant under the transformation $n \to \varphi(n)$, so $\gamma'_{WO}$ is either invariant, or it is transformed into $\gamma''_{WO}$. In order to show that $\gamma'_{WO}$ is invariant, it suffices to show for one pair of points with $c < d$, that $\varphi(c) < \varphi(d)$; here $<$ is the order relation in $\gamma'_{WO}$. To do this, we shall prove a lemma.

**Lemma.** If $\Phi(a, m, p) = 0$ and $\Phi(a, m, 0) = z$, then $pz \, \sigma \, OW$.

PROOF of the lemma. As above, let $R$ be the intersection of the line $y = \Phi(x, m, 0)$ with $XY$. Then, by section 3.1, (8),

$$z = (OR \cap aY)X \cap OW,$$

$$O = \{R(Xp \cap OY) \cap aY\}X \cap OW.$$

The latter provides the construction for $p$:

$$p = \{(aY \cap OX)R \cap OY\}X \cap OW.$$

Now, if $OR \cap aY = S$, we consider the quadrangle $OXYS$. $OX \cap aY = U$, $OY \cap RU = V$, $XS \cap RU = T$. Applying Th. 7.2.3, we see that $UR \, \sigma \, VT$. Projecting from $X$ on $OW$, we obtain $OW \, \sigma \, pz$.

To finish the proof of Th. 7.3.1, we observe that $\varphi(0) = z$, $\varphi(p) = 0$; moreover, by the lemma, if $0 < z$, then $p < 0$, and if $z < 0$, then $0 < p$; so in both cases $\gamma'_{WO}$ is invariant under $\varphi$.

COROLLARY 7.3.1. For $m = 1$ we get: If $n < n'$, then

$$x + n < x + n'.$$

**Theorem 7.3.2.** In $\mathfrak{P}(C1 - C6)$: If $x < x'$, then $x + n < x' + n$.

PROOF. First of all, if, for a fixed $n$, $\psi(x) = x + n$, then $\psi$ is a projectivity from $OW$ onto itself (section 3.2, after (9)), so it leaves the cyclical order $\sigma$ invariant. Moreover, $\psi(W) = W$.

As in the preceding proof, in order to show that $\gamma'_{WO}$ is invariant, it suffices to find $a$ and $b$, such that $a < b$ and $\psi(a) < \psi(b)$. Let $q$ be the solution of $q + n = 0$; the construction of $q$ is as follows: $Xn \cap OY = N$; $WN \cap OX = L$; $YL \cap OW = q$. Applying Th. 7.2.3 to the quadrangle $XYLN$, we see that $qn \, \sigma \, OW$; thus, if $0 < n$, then $q < 0$ and $\psi(q) < \psi(0)$; if $n < 0$, then $0 < q$ and $\psi(0) < \psi(q)$.

In order to exploit completely the invariance of the cyclical order under projection, we study the intersection of the pencil of lines through a fixed point $S = (s, t)$ with a fixed line $l$ ($x = p$) through $Y$. A line $a$ through $S$ is determined by its first coor-

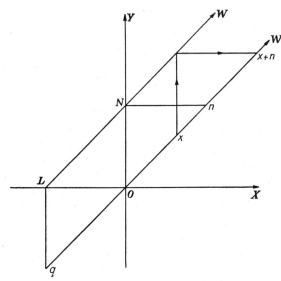

Fig. 7.5.

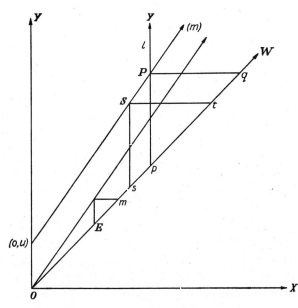

Fig. 7.6.

dinate $m$, for by section 3.1 (vii), $\Phi(s, m, u) = t$ has a unique solution for $u$. Let $a$ intersect $l$ in $P = (p, q)$; then $q$ is a function of $m$, $q = \varphi(m)$.

Given $m$, we construct $q$ by consecutive projections $\pi_1$, $\pi_2$, $\pi_3$, $\pi_4$, as follows:

$\pi_1$ is the projection from $X$ onto $YE$;
$\pi_2$ is the projection from $O$ onto $XY$;
$\pi_3$ is the projection from $S$ onto $l$;
$\pi_4$ is the projection from $X$ onto $OW$.

It follows that the cyclical order $\sigma$ is invariant under $\varphi$. Moreover, $\varphi(W) = W$.

In order to know whether $\gamma'_{WO}$ is invariant or not, we shall investigate the order relation between $\varphi(0)$ and $\varphi(1)$.

$\varphi(0) = t$.

$\varphi(1) = \Phi(p, 1, n)$, where $\Phi(s, 1, n) = t$; that is: $\varphi(1) = p + n$, where $s + n = t$.

By Th. 7.3.2 we have; if $p > s$, then $\varphi(1) > t = \varphi(0)$; if $p < s$, then $\varphi(1) < t = \varphi(0)$.

Thus, if $p > s$, $\gamma'_{WO}$ is invariant under $\varphi$; if $p < s$, $\gamma'_{WO}$ is transformed by $\varphi$ into $\gamma''_{WO}$. We formulate this result in the following theorem:

**Theorem 7.3.3.** Let the lines $[m_1, n_1]$, $[m_2, n_2]$ intersect in $S = (s, t)$. Let $m_1 < m_2$.

If $x > s$, then $\Phi(x, m_1, n_1) < \Phi(x, m_2, n_2)$;
if $x < s$, then $\Phi(x, m_1, n_1) > \Phi(x, m_2, n_2)$.

The properties of the ternary field, which are contained in Th. 7.3.3, can be used for the definition of an ordered ternary field.

**Definition.** A ternary field $\mathfrak{T}$ is an *ordered ternary field*, if it is ordered by a relation $<$, for which the following properties (1) and (2) hold.

(1)　　　　If $n_1 < n_2$, then $\Phi(x, m, n_1) < \Phi(x, m, n_2)$.

(2)　　　　If $\Phi(s, m_1, n_1) = \Phi(s, m_2, n_2)$,

then for $p > s$ and $m_1 < m_2$, we have $\Phi(p, m_1, n_1) < \Phi(p, m_2, n_2)$;
and for $p < s$ and $m_1 < m_2$, we have $\Phi(p, m_1, n_1) > \Phi(p, m_2, n_2)$.

Theorems 7.3.1 and 7.3.3 can now be summarized as follows: In $\mathfrak{P}$ (C1—C6), $\mathfrak{T}_\mathfrak{P}$ is made into an ordered ternary field by the order $\gamma'_{WO}$.

We shall now derive some simple properties of ordered ternary fields.

**Theorem 7.3.4.** In every ordered ternary field, if $m_1 < m_2$ and $x > 0$, then $\Phi(x, m_1, n) < \Phi(x, m_2, n)$; if $m_1 < m_2$ and $x < 0$, then $\Phi(x, m_1, n) > \Phi(x, m_2, n)$.

PROOF. In (2) we take $s = 0$; $\Phi(0, m_1, n) = \Phi(0, m_2, n) = n,$. so $n_1 = n_2$. This gives the desired result.

As a special case ($n = 0$) we have:

COROLLARY 7.3.4. If $m_1 < m_2$ and $x > 0$, then $xm_1 < xm_2$; if $m_1 < m_2$ and $x < 0$, then $xm_1 > xm_2$.

**Theorem 7.3.5.** In every ordered ternary field, if $x_1 < x_2$ and $m > 0$, then $\Phi(x_1, m, n) < \Phi(x_2, m, n)$; if $x_1 < x_2$ and $m < 0$, then $\Phi(x_1, m, n) > \Phi(x_2, m, n)$.

PROOF. In (2) we put $m_1 = 0$, $s = x_1$, $p = x_2$, $m_2 = m$, $n_2 = n$. As $\Phi(x_1, 0\ n_1) = \Phi(x_2, 0, n_1) = n_1$, we obtain: If $n_1 = \Phi(x_1, m, n)$ then for $x_2 > x_1$ and $0 < m$ we have $n_1 < \Phi(x_2, m, n)$; in other words: If $x_1 < x_2$ and $m > 0$, then $\Phi(x_1, m, n) < \Phi(x_2, m, n)$. The second part of the theorem follows in the same way.

COROLLARY 7.3.5. 1. If $m > 0$ and $x_1 < x_2$, then $x_1m < x_2m$; if $m < 0$ and $x_1 < x_2$, then $x_1m > x_2m$.

COROLLARY 7.3.5. 2. If $x_1 < x_2$, then $x_1 + n < x_2 + n$.

COROLLARY 7.3.5. 3. An ordered ternary field contains an infinite number of elements (apply corollary 2 with $x_1 = 0$, $x_2 = 1$, $n = 1$, and repeat this procedure).

**Theorem 7.3.6.** In every ordered ternary field, if $\Phi(s, m_1, n_1) = \Phi(s, m_2, n_2)$ and $\Phi(p, m_1, n_1) < \Phi(p, m_2, n_2)$, then either $p > s$ and $m_1 < m_2$, or $p < s$ and $m_1 > m_2$.

PROOF. Other possibilities are excluded by (2).

## § 7.4. The geometry over an ordered ternary field.

Let $\mathfrak{T}$ be an ordered ternary field. As we have just seen, the following properties hold in $\mathfrak{T}$.

(1)   If $n_1 < n_2$, then $\Phi(x, m, n_1) < \Phi(x, m, n_2)$, and in particular, $x + n_1 < x + n_2$.

(2a)  If $\Phi(s, m_1, n_1) = \Phi(s, m_2, n_2)$, $p > s$ and $m_1 < m_2$, then $\Phi(p, m_1, n_1) < \Phi(p, m_2, n_2)$.

(2b)  If $\Phi(s, m_1, n_1) = \Phi(s, m_2, n_2)$, $p < s$ and $m_1 < m_2$, then $\Phi(p, m_1, n_1) > \Phi(p, m_2, n_2)$.

(2c)  If $\Phi(s, m_1, n_1) = \Phi(s, m_2, n_2)$ and $\Phi(p, m_1, n_1) < \Phi(p, m_2, n_2)$, then either $p > s$ and $m_1 < m_2$, or $p < s$ and $m_1 > m_2$.

(3a)  If $m_1 < m_2$ and $x > 0$, then $\Phi(x, m_1, n) < \Phi(x, m_2, n)$, and in particular, $x m_1 < x m_2$.

(3b)  If $m_1 < m_2$ and $x < 0$, then $\Phi(x, m_1, n) > \Phi(x, m_2, n)$, and in particular, $x m_1 > x m_2$.

(4a)  If $x_1 < x_2$ and $m > 0$, then $\Phi(x_1, m, n) < \Phi(x_2, m, n)$, and in particular, $x_1 m < x_2 m$.

(4b)  If $x_1 < x_2$ and $m < 0$, then $\Phi(x_1, m, n) > \Phi(x_2, m, n)$, and in particular, $x_1 m > x_2 m$.

(5)   If $x_1 < x_2$, then $x_1 + n < x_2 + n$.

Now let $\mathfrak{P}(\mathfrak{T})$ be the projective plane over $\mathfrak{T}$ as defined in section 3.4. $\mathfrak{P}(\mathfrak{T})$ contains the line $XY = \omega$; $\mathfrak{P} \setminus \omega$ is called the affine plane.

By Th. 3.4.1, the ternary field attached to $\mathfrak{P}(\mathfrak{T})$ by the coordinate system $OXYE$ and consisting of the points of $OW \setminus \{W\}$, is isomorphic to $\mathfrak{T}$. We identify corresponding elements in this isomorphism, so that $\mathfrak{T}$ is the ternary field attached to $\mathfrak{P}(\mathfrak{T})$.

Now, as $\mathfrak{T}$ is ordered, $OW \setminus \{W\}$ is an ordered set. We extend this order to any line not containing $Y$ by projection from $Y$; in other words, if $A = (a, b)$ and $C = (c, d)$ are points such that $AC$ does not contain $Y$, we define $A < C$ by $a < c$. If is clear that by this definition we obtain an order relation on every line not through $Y$. This order relation is only defined for the points of the affine plane.

If $l$ does not contain $X$, we can define another order relation on $l$ by projection from $X$, namely if $A = (a, b)$ and $C = (c, d)$, then $A \overset{\cdot}{<} C$ is defined by $b < d$.

**Theorem 7.4.1.** If $l$ contains neither $X$ nor $Y$, then either for any two points $A$, $C$ of $l$, $A < C$ implies $A \overset{\cdot}{<} C$, or, for any two points of $l$, $A < C$ implies $C \overset{\cdot}{<} A$.

PROOF. Let $A = (a, b)$, $C = (c, d)$, and let $y = \Phi(x, m, n)$ be the equation of $l$. By (4a) and (4b) above,

if $m > 0$, then $A < C$ (i.e. $a < c$) implies $b < d$ (i.e. $A \overset{\cdot}{<} C$);

if $m < 0$, then $A < C$, (i.e. $a < c$) implies $d < b$ (i.e. $C \overset{\cdot}{<} A$).

On a line through $X$ only the relation $<$ is defined, on a line through $Y$ only the relation $\overset{\cdot}{<}$. For $\omega$ no order relation has been defined so far.

We extend the order relations on $l$ by adding the point of intersection of $l$ with $\omega$ as the first point. As was proved in Th. 7.1.8, alternating pairs on $l$ define a cyclical order of $l$.

**Theorem 7.4.2.** If the order relations $<$ and $\overset{\cdot}{<}$ are both defined on $l$, they define the same cyclical order.

PROOF. Suppose that $A = (a, b)$, $C = (c, d)$, $E = (e, f)$ and $G = (g, h)$ are given on $l$ such that $AE$ and $CG$ are alternating pairs for the relation $<$; for instance, $A < C < E < G$; then $a < c < e < g$. By the preceding theorem, either $b < d < f < h$ or $h < f < d < b$. In both cases $AE$ and $CG$ are alternating pairs for the relation $\overset{\cdot}{<}$.

Thus we obtain one cyclical order for every line, except $\omega$, which we denote by $\sigma$.

We extend it to $\omega$ as follows. The lines through a given point $M$ of $\omega$ have the same first coordinate $m$; we denote, as before, $M$ by $(m)$.

**Definition.** The order relation $<$ on $\omega$ is defined by $(m_1) < (m_2)$ if $m_1 < m_2$; $Y$ is added as the first element. The cyclical order $\sigma$ on $\omega$ is the relation between alternating pairs with respect to $<$.

We wish to prove that the cyclical order $\sigma$, which is now defined on every line, satisfies C1—C6.

This is clear for C1—C4. C5 follows from Corollary 3 to Th.

7.3.5; C6 will be proved in a series of theorems, which ends in Th. 7.4.12.

**Theorem 7.4.3.** The cyclical order $\sigma$ is invariant under projection from $X$ as well as from $Y$.

PROOF. This is clear from the definitions of $<$ and $\dot{<}$.

**Theorem 7.4.4.** The cyclical order $\sigma$ is invariant under projection from any point of $XY$.

PROOF. Let $R$ be a point of $XY$; the first coordinate of every line through $R$ has the same value $m$. Let $l_1 = [m, n_1]$ and $l_2 = [m, n_2]$ be two lines through $R$ such that $n_1 < n_2$, and let $l^* = [m^*, n^*]$ be any line not through $R$.

$l^*$ intersects $l_1$ in $P_1 = (p_1, q_1)$, $l_2$ in $P_2 = (p_2, q_2)$, so that $q_1 = \Phi(p_1, m, n_1) = \Phi(p_1, m^*, n^*)$ and $q_2 = \Phi(p_2, m, n_2) = \Phi(p_2, m^*, n^*)$.

$\Phi(p_1, m, n_1) < \Phi(p_1, m, n_2)$ by (1). Thus we have $\Phi(p_2, m^*, n^*) = \Phi(p_2, m, n_2)$ and $\Phi(p_1, m^*, n^*) < \Phi(p_1, m, n_2)$. It follows from (2c) that the last two relations are only compatible if either

$$p_1 > p_2 \text{ and } m^* < m, \text{ or } p_1 < p_2 \text{ and } m^* > m.$$

Consequently, if $m^* > m$, the order relation of $P_1$ and $P_2$ is the same as that of $n_1$ and $n_2$; if $m^* < m$, it is the opposite. In both cases alternating pairs of points on $l^*$ correspond to alternating pairs of values of $n$. It follows that the cyclical order is invariant under projection from $R$.

**Theorem 7.4.5.** Let $ABC$ be a triangle in the affine plane; let $BC, CA, AB$ intersect $\omega$ in $M, M', M''$ and a line $l$, not through $A, B, C$, in $P, P', P''$ respectively. Then $BC \, \sigma \, PM$ entails that either $CA \, \sigma \, P'M'$ or $AB \, \sigma \, P''M''$, but not both.

PROOF. It suffices to prove that

(I)    $BC \, \sigma \, PM$ and $CA \, \rho \, P'M'$ entail $AB \, \sigma \, P''M''$,
(II)   $BC \, \sigma \, PM$ and $CA \, \sigma \, P'M'$ entail $AB \, \rho \, P''M''$.

(I). Put $l \cap \omega = M^*$, $M^*C \cap AB = Q$. Using the preceding theorem we find, by projection from $M^*$,

$BC \, \sigma \, PM$ entails $BQ \, \sigma \, P''M''$; $CA \, \rho \, P'M'$ entails $QA \, \rho \, P''M''$.
Th. 7.1.2 gives the desired result $AB \, \sigma \, P''M''$.

(II) is proved analogously.

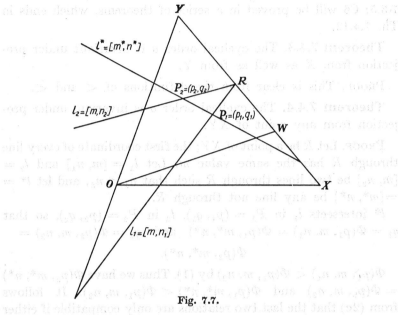

Fig. 7.7.

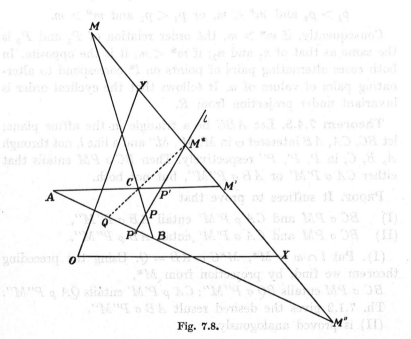

Fig. 7.8.

It is left to the reader to verify that (I) and (II) are easy to prove in the special cases where $P = M$, $P' = M'$ or (and) $P'' = M''$, provided the hypotheses of the theorem are satisfied.

REMARK. In the affine plane, $BC \, \sigma \, PM$ can be read as "$P$ is between $B$ and $C$" or "$P$ is on the side $BC$ of the triangle $ABC$". Then Th. 7.4.5 becomes: If a line $l$ intersects a side of a triangle, and if $l$ contains no vertex of that triangle, then it intersects exactly one other side of the triangle. In this form it is known as Pasch's theorem. However, we prefer the version above.

**Definition.** Let $l$ be a line and let $A$, $B$ be distinct points in $\mathfrak{P} \setminus (l \cup \omega)$, where $AB$ intersects $l$ in $P$, $\omega$ in $M$; then $\delta_l \, (A, B)$ means $AB \, \rho \, PM$. $\delta_l(A, A)$ is always true if $A \, \epsilon \, \mathfrak{P} \setminus (l \cup \omega)$.

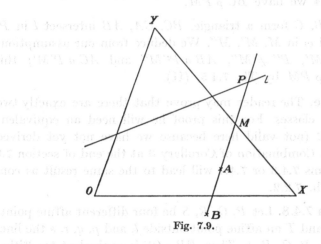

Fig. 7.9.

$\delta_l(A, B)$ can be read as "$A$ and $B$ are on the same side of $l$". This interpretation is justified by the following two theorems.

**Theorem 7.4.6.** For a given line $l$, $\delta_l$ is an equivalence relation on $\mathfrak{P} \setminus (l \cup \omega)$.

PROOF. We have to prove that $\delta_l$ is reflexive, symmetric and transitive. The first two properties are obvious. Transitivity means that $\delta_l(A, B)$ and $\delta_l(B, C)$ entail $\delta_l(A, C)$. Two cases must be considered:

(I). $A$, $B$, $C$ are collinear. $AB \cap l = P$, $AB \cap \omega = M$, then $AB \, \rho \, PM$, $BC \, \rho \, PM$ entail $AC \, \rho \, PM$ by Th. 7.1.3.

(II). $A$, $B$, $C$ are not collinear. $BC$, $CA$, $AB$ intersect $l$ in $P$, $P'$, $P''$, $\omega$ in $M$, $M'$, $M''$. It follows immediately from Th. 7.4.5 (I) that $AB \rho P''M''$, $BC \rho PM$ imply $AC \rho P'M'$.

**Theorem 7.4.7.** There are at most two equivalence classes in $\mathfrak{P} \setminus (l \cup \omega)$ with respect to the relation $\delta_l$.

PROOF. Let us assume that $A$ and $B$ are in different classes, and that $A$ and $C$ are in different classes; we must prove that $B$ and $C$ are in the same class. As in the proof of Th. 7.4.6 there are two cases:

(I) $A$, $B$, $C$ collinear: $AB \cap l = P$, $AB \cap \omega = M$. From our assumptions it follows that $P \neq M$, $AB \sigma PM$ and $AC \sigma PM$; by Th. 7.1.4 we have $BC \rho PM$.

(II). $A$, $B$, $C$ form a triangle. $BC$, $CA$, $AB$ intersect $l$ in $P$, $P'$, $P''$, and $\omega$ in $M$, $M'$, $M''$. We deduce from our assumptions that $P' \neq M'$, $P'' \neq M''$, $AB \sigma P''M''$ and $AC \sigma P'M'$; this entails $BC \rho PM$ by Th. 7.4.5, (II).

**Excercise.** The reader may prove that there are exactly two equivalence classes. For this proof he will need an equivalent of Th. 7.2.2 (not valid here because we have not yet derived C5 and C6). Combination of Corollary 3 at the end of section 7.3 with theorems 7.4.3 or 7.4.4 will lead to the same result as contained in Th. 7.2.2.

**Theorem 7.4.8.** Let $P$, $Q$, $R$, $S$ be four different affine points on a line $l$, and $T$ an affine point outside $l$, and $p$, $q$, $r$, $s$ the lines joining $T$ to $P$, $Q$, $R$, $S$. Then $PR \sigma QS$ is equivalent to: Either $\delta_q(P, R)$ is true and $\delta_s(P, R)$ is false, or $\delta_q(P, R)$ is false and $\delta_s(P, R)$ is true.

PROOF. Let $l$ intersect $\omega$ in $M$.

Suppose $PR \sigma QS$ and $\delta_q(P, R)$, i.e. $PR \rho QM$; then $PR \sigma SM$, i.e. not $\delta_s(P, R)$.

Also, if $PR \sigma QS$ and not $\delta_q(P, R)$ we have $PR \sigma QM$, which gives $PR \rho SM$, i.e. $\delta_s(P, R)$.

Conversely, if $\delta_q(P, R)$ and not $\delta_s(P, R)$, we have $PR \rho QM$ and $PR \sigma SM$; it follows that $PR \sigma QS$.

If $\delta_s(P, R)$ and not $\delta_q(P, R)$, the proof is analogous.

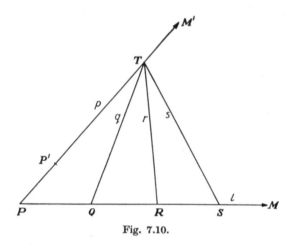

Fig. 7.10.

REMARK. The theorem remains valid if either $Q$ or $S$ is on $\omega$. For instance, if $Q \in \omega$, then $\delta_q(P, R)$ is true and $PR\,\sigma\,QS$ is equivalent to not $\delta_s(P, R)$.

In order to express the preceding theorem more conveniently, the relation $\delta_{qs}(P, R)$ is defined as follows:

**Definition.** In $\mathfrak{P} \setminus (q \cup s \cup \omega)$, $\delta_{qs}(P, R)$ means: Either $\delta_q(P, R)$ is true and $\delta_s(P, R)$ is false, or $\delta_q(P, R)$ is false and $\delta_s(P, R)$ is true.

Then Th. 7.4.8. becomes:

**Theorem 7.4.9.** Let $l$, $P$, $Q$, $R$, $S$, $T$, $p$, $q$, $r$, $s$ be as in Th. 7.4.8. Then $PR\,\sigma\,QS$ is equivalent to $\delta_{qs}(P, R)$.

**Theorem 7.4.10.** Let $l$, $P$, $Q$, $R$, $S$, $T$, $p$, $q$, $r$, $s$ be as in Th. 7.4.8; let $P'$ be an affine point on $p$, different from $T$. Then $\delta_{qs}(P,R)$ entails $\delta_{qs}(P', R)$.

PROOF. Let $p$ intersect $\omega$ in $M'$. $\delta_q(P, P')$ and $\delta_s(P,P')$ are equivalent to $PP'\,\rho\,TM'$.

Thus, if $\delta_q(P, P')$, it follows from $\delta_q(P, R)$ (by Th. 7.4.6) that $\delta_q(P', R)$, and from not $\delta_s(P, R)$ that not $\delta_s(P', R)$.

If not $\delta_q(P, P')$, then it follows from $\delta_q(P, R)$ that not $\delta_q(P', R)$, and from not $\delta_s(P, R)$ that $\delta_s(P', R)$ (by theorems 7.4.6 and 7.4.7).

Thus in both cases it follows from $\delta_{qs}(P, R)$ that $\delta_{qs}(P', R)$.

**Theorem 7.4.11.** Let $p$, $q$, $r$, $s$ be four lines through $T$, which intersect $l$ in $P$, $Q$, $R$, $S$ and $l'$ in $P'$, $Q'$, $R'$, $S'$, while none of these points is on $\omega$. Then we have: If $PR\ \sigma\ QS$, then $P'R'\ \sigma\ Q'S'$.

PROOF. If $PR\ \sigma\ QS$, then by Th. 7.4.9, $\delta_{qs}(P, R)$; by Th. 7.4.10, applied twice, $\delta_{qs}(P', R')$; by Th. 7.4.9 this is equivalent to $P'R'\ \sigma\ Q'S'$.

Th. 7.4.11 expresses the invariance of the cyclical order under projection for the case in which all the points under consideration are in the affine plane. It must still be extended to the case in which some of the points are on $\omega$.

The case $T \in \omega$ has been treated in Th. 7.4.4.

If at most one of the points $Q$, $S$ and at most one of the points $Q'$, $S'$ are on $\omega$, the proof given above remains valid by the remark following Th. 7.4.8.

If at most one of $P$, $R$ and at most one of $P'$, $R'$ are on $\omega$, we can apply the proof after interchanging $P$, $R$ with $Q$, $S$.

If, for instance, $P$ and $Q'$ are on $\omega$, we insert in the proof an auxiliary line which contains neither $T$ nor $P$ nor $Q'$.

The only case which remains to be considered is the one where $l$ or $l'$ coincides with $\omega$. Let the lines $\lceil m_1, n_1 \rceil$ and $[m_2, n_2]$ intersect in $S = (s, t)$. As a special case, for $p = 0$, of property (2) of an ordered ternary field (see the beginning of § 7.4) we have: If $s < 0$ and $m_1 < m_2$, then $n_1 < n_2$; if $s > 0$ and $m_1 < m_2$, then $n_1 > n_2$. Thus by projection of $\omega$ from $S$ onto $OY$, the order relation is either preserved or inversed; in both cases the cyclical order $\sigma$ is invariant. By the preceding theorem it is then invariant under projection on any line.

We have now attained the announced result: Axiom C6 is valid for $\sigma$. Thus we have:

**Theorem 7.4.12.** If $\mathfrak{T}$ is an ordered ternary field, it is possible to introduce a cyclical order $\sigma$ in $\mathfrak{P}(\mathfrak{T})$, such that $\sigma$ satisfies axioms C1—C6 and that on $OW$ $\sigma$ coincides with the cyclical order $\tau$, defined by the relation of alternating pairs in the order of $\mathfrak{T}$.

The case where $D_{10}$ is valid in $\mathfrak{P}$ deserves some attention. (As we noted in Cor. 2 to Th. 7.2.3, $D_9$ and $D_{10}$ are equivalent in $\mathfrak{P}(C1—C6)$.) In $\mathfrak{T}(D_{10})$ the properties (i)—(vii) mentioned in

Th. 3.4.2 hold, therefore the definition of an ordered ternary field can be much simplified in this case.

**Theorem 7.4.13.** A necessary and sufficient condition that $\mathfrak{T}(D_{10})$ be an ordered ternary field with the order relation $<$, is that the following properties $(\alpha)$, $(\beta)$ hold universally in $\mathfrak{T}(D_{10})$:

$(\alpha)$   if $a < b$, then $c + a < c + b$.

$(\beta)$   if $a > 0$ and $b > 0$, then $ab > 0$.

PROOF. The necessity of these conditions follows immediately from properties (1) and (4a) of an ordered ternary field.

Now let $(\alpha)$ and $(\beta)$ be valid in $\mathfrak{T}(D_{10})$. We must prove the properties (1) and (2) occurring in the definition of an ordered ternary field. In this case (1) and (2) become

(1')   If $n_1 < n_2$, then $xm + n_1 < xm + n_2$.

(2')   If $sm_1 + n_1 = sm_2 + n_2$, $p > s$ and $m_1 < m_2$,     then

$pm_1 + n_1 < pm_2 + n_2$.

If $sm_1 + n_1 = sm_2 + n_2$, $p < s$ and $m_1 < m_2$,     then

$pm_1 + n_1 > pm_2 + n_2$.

(1') is a special case of $(\alpha)$.

To prove the first part of (2'), we solve $sm_1 + n_1 = sm_2 + n_2$ for $n_2$ and substitute the result in the inequality to be proved. By means of (i)—(vii), it is easily reduced to $(p-s)(m_2-m_1) > 0$, which is true on the basis of $(\alpha)$ and $(\beta)$.

The second part of (2') is proved similarly.

It can be proved that $\mathfrak{T}(D_{10})$, if it is ordered, is a division ring. In other words, $D_{11}$ follows from V1—V3, $D_{10}$, C1—C6. We shall not give the proof, because it is entirely algebraical.

[Bruck & Kleinfeld, Proc. Amer. Math. Soc. **2** (1951), p. 887. The theorem follows immediately from Theorem 5 of Skorniakow, Ukrainskiĭ Mat. Žurnal **2**, part 1 (1950), p. 81.]

**Definition.** A division ring in which an order relation satisfying $(\alpha)$ and $(\beta)$ is defined, is called an *ordered division ring*.

It follows easily from Cor. 2 to Th. 7.3.5, that every ordered division ring has characteristic 0.

## § 7.5. A counterexample.

Hilbert was the first to give an example of an ordered non-commutative division ring. Before we can explain his example, some preparation is necessary.

**Formal power series.** Let $L$ be a field and $t$ a variable (or, in algebraic language, let $t$ be transcendent with respect to $L$). The field $L^*(t)$ of formal power series over $L$ is constructed as follows. The elements of $L^*(t)$ have the form

$$(1) \qquad \alpha = \alpha_k t^k + \alpha_{k+1} t^{k+1} + \cdots,$$

$k$ an integer $\geqq 0$, $\alpha_i \epsilon L$ $(i = k, \ldots)$, $\alpha_k \neq 0$. We can also write

$$(2) \quad \alpha = \sum_{i=-\infty}^{+\infty} \alpha_i t^i, \text{ where there is associated with } \alpha \text{ an integer}$$

$k = k(\alpha)$, such that $\alpha_i = 0$ for $i < k(\alpha)$, $\alpha_i \neq 0$ for $i = k(\alpha)$. A zero element 0 is added to $L^*(t)$.

Addition and multiplication are defined as follows.

$$\text{If } \alpha = \sum_{i=-\infty}^{+\infty} \alpha_i t^i, \ \beta = \sum_{i=-\infty}^{+\infty} \beta_i t^i, \text{ then}$$

$$(3) \qquad \alpha + \beta = \Sigma(\alpha_i + \beta_i)t^i.$$

Evidently, $k(\alpha + \beta) \geqq \min \ (k(\alpha), \ k(\beta))$.
$\gamma = \alpha\beta$ means that $\gamma = \Sigma \gamma_i t^i$ with

$$(4) \qquad \gamma_i = \sum_{h+j=i} \alpha_h \beta_j.$$

(Verify that only a finite number of terms in this sum are not zero).

$$k(\alpha\beta) = k(\alpha) + k(\beta).$$

It is easily verified that $L^*(t)$ is an abelian group with respect to addition, and that the commutative and associative laws for multiplication hold, as well as the distributive law. We shall prove the existence of $\alpha^{-1}$ for $\alpha \neq 0$. Let us try to find $\beta$ such that $\alpha\beta = 1$. It is clear that $k(\beta) = -k(\alpha)$ and that $\beta_{k(\beta)} = (\alpha_{k(\alpha)})^{-1}$. Let $\beta_j$ be already known for $k(\beta) \leq j \leq m$.

Consider the equation (4) for $i = k(\alpha) + m + 1$; it becomes

$$(5) \quad 0 = \alpha_{k(\alpha)} \beta_{m+1} + \alpha_{k(\alpha)+1}\beta_m + \cdots + \alpha_{k(\alpha)+m+1-k(\beta)}\beta_{k(\beta)},$$

which allows us to calculate $\beta_{m+1}$.

In this way the coefficients of $\beta$ can be calculated one after the other.

If $L$ is an ordered field, $L^*(t)$ can be ordered by the following rules.

(i)       $\alpha > 0$ if and only if $\alpha_{k(\alpha)} > 0$.

(ii)      $\alpha > \beta$ if and only if $\alpha - \beta > 0$.

It is easy to verify that (i) and (ii) define an order relation in $L^*(t)$ and, moreover, that $L^*(t)$ is an ordered field with respect to this order relation; that is, if $\alpha > \beta$, then $\alpha + \gamma > \beta + \gamma$, and if $\alpha > 0$ and $\beta > 0$, then $\alpha\beta > 0$.

We now define a mapping $\varphi$ of $L^*(t)$ onto itself, as follows. Let $c$ be an element of $L$, $c > 0$.

(6)    If $\alpha = \sum_{i=-\infty}^{+\infty} \alpha_i t^i$, then $\varphi(\alpha) = \sum_{i=-\infty}^{+\infty} c^i \alpha_i t^i$.

**Theorem 7.5.1.** The mapping $\varphi$, defined by (6), is an automorphism of $L^*(t)$.

PROOF. It is easy to see that $\varphi$ is one-to-one.

It must be verified that $\varphi(\alpha + \beta) = \varphi(\alpha) + \varphi(\beta)$ and that $\varphi(\alpha\beta) = \varphi(\alpha)\varphi(\beta)$. The former is obvious, because $c^i(\alpha_i + \beta_i) = c^i\alpha_i + c^i\beta_i$. As to the latter, put $\alpha\beta = \gamma$, $\varphi(\alpha)\varphi(\beta) = \delta$. By (4),

$$\delta_i = \sum_{h+j=i} c^h\alpha_h\, c^j\beta_j = c^i \sum_{h+j=i} \alpha_h\beta_j = c^i\gamma_i.$$

**Theorem 7.5.2.** If $L$ is an ordered field and $c > 0$, then the mapping $\varphi$, defined by (6), preserves the order relations in $L^*(t)$, as defined by (i) and (ii).

PROOF. Clear, for if $\alpha_{k(\alpha)} > 0$, then $c^{k(\alpha)} \alpha_{k(\alpha)} > 0$.

Now we come to Hilbert's example.

Let $K = R^*(t)$ be the field of formal power series in $t$ over the field $R$ of rationals, ordered as defined above for $L^*(t)$. After choosing a rational number $c > 0$ $(c \neq 1)$, we consider the automorphism $\varphi$ defined by (6).

We could form the field $K^*(s)$ of formal power series over $K$; however, instead of $K^*(s)$, we consider a division ring $K^\dagger(s)$, which differs from $K^*(s)$ only in the definition of multiplication.

Thus the elements of $K^\dagger(s)$ are the formal power series over $K$:

$$A = \sum_{i=-\infty}^{+\infty} A_i s^i; \quad A_i \in K; \quad A_i = 0 \text{ for } i < k(A); \quad A_{k(A)} \neq 0.$$

If $B = \sum_{i=-\infty}^{+\infty} B_i s^i$, then

(7) $$A + B = \sum(A_i + B_i)s^i.$$

(8) $$AB = \sum D_j s^j \text{ with } D_j = \sum_{h+i=j} A_i \varphi^i(B_h).$$

The order relation in $K^\dagger(s)$ is defined in the same way as in $L^*(t)$, namely

(i) $\qquad\qquad A > 0$ if and only if $A_{k(A)} > 0$.

(ii) $\qquad\qquad A > B$ if and only if $A - B > 0$.

**Theorem 7.5.3.** $K^\dagger(s)$ is an ordered division ring.

PROOF. The following properties need to be considered; the others are trivial.

(i)   Multiplication is associative.

(ii)  Multiplication is distributive with respect to addition.

(iii) Every element $\neq 0$ has an inverse.

(iv)  If $A > 0$ and $B > 0$, then $AB > 0$.

(i). Let $A = \sum A_i s^i$, $B = \sum B_i s^i$, $C = \sum C_i s^i$ be given. If $AB = D$, $DC = E$, $BC = F$, $AF = G$, we must prove that $E = G$.

$$D_h = \sum_{i+j=h} A_i \varphi^i B_j.$$

$$E_k = \sum_{h+l=k} D_h \varphi^h C_l = \sum_{i+j+l=k} A_i \varphi^i B_j \varphi^{i+j} C_l.$$

$$F_h = \sum_{j+l=h} B_j \varphi^j C_l.$$

$$G_k = \sum_{i+h=k} A_i \varphi^i F_h = \sum_{i+j+l=k} A_i \varphi^i (B_j \varphi^j C_l) =$$
$$= \sum_{i+j+l=k} A_i \varphi^i B_j \varphi^{i+j} C_l.$$

In the last reduction the fact that $\varphi$ is an isomorphism has been utilized. As the results for $E_k$ and $G_k$ are identical, we have $E = G$.

(ii). Let $A$, $B$, $C$ be given as under (1).

If $AB = D$, $AC = E$, $D + E = F$, $B + C = G$, $AG = H$, we must prove that $F = H$.

$$D_h = \sum_{i+j=h} A_i \varphi^i B_j. \quad E_h = \sum_{i+j=h} A_i \varphi^i C_j.$$

$$F_h = \sum_{i+j=h} (A_i \varphi^i B_j + A_i \varphi^i C_j) =$$

$$= \sum_{i+j=h} A_i (\varphi^i B_j + \varphi^i C_j)$$

$$= \sum_{i+j=h} A_i \varphi^i (B_j + C_j).$$

$$G_j = B_j + C_j.$$

$$H_h = \sum_{i+j=h} A_i \varphi^i G_j = \sum_{i+j=h} A_i \varphi^i (B_j + C_j).$$

As the results for $F_h$ and $H_h$ are identical, we have $F = H$. The proof of the second distributive property, viz.

$$BA + CA = (B + C) A, \text{ is left to the reader.}$$

(iii). The calculation of $A^{-1}$ is analogous to the case of $K^*(t)$. We should find $H$ such that $AH = 1$. First of all, $k(H) = -k(A)$ and $H_{k(H)} = (A_{k(A)})^{-1}$.

Let $H_j$ be already known for $k(H) \leq j \leq m$.

Equation (8) for $j = k(A) + m + 1$ becomes

(9) $\quad 0 = A_{k(A)} \varphi^{k(A)} H_{m+1} + A_{k(A)+1} \varphi^{k(A)+1} H_m + \ldots + A_l \varphi^l H_{k(H)},$

where $l = 2k(A) + m + 1$.

From (9) we can find $H_{m+1}$, so that the coefficients of $H$ become known consecutively.

(iv). If $A > 0$ and $B > 0$, then $A_{k(A)} > 0$ and $B_{k(B)} > 0$.

$$AB = C \text{ entails } C_{k(C)} = A_{k(A)} \varphi^{k(A)} B_{k(B)}.$$

By the order preserving property of $\varphi$, $\varphi^{k(A)} B_{k(B)} > 0$; and because $K$ is an ordered field, $C_{k(C)} > 0$, that is $C > 0$.

In the projective plane $\mathfrak{P}^2(K^\dagger(s))$ constructed over Hilbert's ordered division ring, axioms V1—V3, $D_{11}$, C1—C6 are valid, but $P_{10}$ is not. Thus we find

**Theorem 7.5.4.** $P_{10}$ is independent of V1—V3, $D_{11}$, $C_1$—$C_6$.

Analogously the space geometry $P^3(K^\dagger(s))$ shows us:

**Theorem 7.5.5.** $P_{10}$ is independent of R1—R5, C1—C6.

## § 7.6. The axiom of Archimedes.

Let $\mathfrak{G}$ be an ordered group in which the group relation is denoted by $+$. If $a$ is an element of $\mathfrak{G}$, we can form the sequence $a, a + a,$ $a + a + a, \ldots$; the elements of this sequence are denoted briefly by $a, 2a, 3a, \ldots$. More exactly, by definition $1a = a$, and for every natural number $n$, $(n + 1)a = na + a$.

(Note that the natural numbers need not be members of $\mathfrak{G}$!)

If $a > 0$, then $(n + 1)a > na$ for every $n$. $\mathfrak{G}$ is called *Archimedean* if it has the following property: For every $a$ and $b$ in $\mathfrak{G}$, where $a > 0$, there exists a natural number $n$ such that $na > b$.

If $D_{10}$ holds in a projective plane $\mathfrak{P}$, then $\mathfrak{T}(\mathfrak{P})$ is a group under addition. Therefore it makes sense to add the following axiom to V1—V3, $D_{10}$, C1—C6.

**Archimedean axiom A.** The additive group of $\mathfrak{T}(\mathfrak{P})$ is Archimedean.

The axiom can be stated in a more geometric form. To this effect we describe geometrically the construction of $(n + 1)a$ from $na$ and $a$; thus, starting with $a$, we obtain a sequence of points on $OW$ for which the axiom can be formulated. However, because the axiom we obtain in this way is nothing else but a translation of the algebraic axiom above, we shall not formulate it explicitly.

Another geometric interpretation will be stated in Th. 7.6.7.

Let us call an ordered division ring Archimedean if its additive group is Archimedean. We begin by deriving some properties of Archimedean ordered division rings.

**Theorem 7.6.1.** In an Archimedean ordered division ring, the field of rationals is dense.

REMARK. Every ordered division ring $\mathfrak{F}$ has the characteristic 0; hence it contains a field $\mathfrak{F}_0$ which is isomorphic to the field $R_0$ of rationals; we identify $\mathfrak{F}_0$ with $R_0$. That $R_0$ is dense in $\mathfrak{F}$ means the following: If $a, b \in \mathfrak{F}$ and $a > b$, then there is an element $r$ of $R_0$ such that $b < r < a$.

PROOF of Th. 7.6.1. Let $a, b$ be elements of $\mathfrak{F}$ such that $a > b > 0$; then $d = a - b > 0$. Because the additive group of $\mathfrak{F}$ is Archimedean we can find a natural number $n$ such that $n \cdot d > 1$

and a natural number $m$ such that $m \cdot 1/n > b$; we may suppose that $m$ is the smallest natural number with this property. Now we have $d > 1/n$ and $(m-1)/n \leq b$, so $m/n \leq b + 1/n < b + d = a$. This gives the desired result, $b < m/n < a$.

**Exercise.** Complete the proof for the case where $a$ and $b$ are not both positive.

**Theorem 7.6.2.** If the ordered division ring $\mathfrak{F}$ contains a subfield $\mathfrak{G}$ which is dense in $\mathfrak{F}$, then $\mathfrak{F}$ is a field.

PROOF. Let $a$, $b$ be elements of $\mathfrak{F}$; suppose that $a > 0$ and $b > 0$ and that $\delta = ab - ba > 0$. We choose an element $c$ of $\mathfrak{F}$ such that $c > \max(a, b)$. It is easily seen that we can find elements $r_1$, $r_2$, $s_1$, $s_2$ of $\mathfrak{G}$ such that

(1) $\qquad 0 < r_1 < a < r_2 < c, \; 0 < s_1 < b < s_2 < c$, and

(2) $\qquad\qquad r_2 - r_1 < \delta/2c, \; s_2 - s_1 < \delta/2c$.

From (1) it follows:

$$0 < r_1 s_1 < ab < r_2 s_2$$
$$0 < r_1 s_1 < ba < r_2 s_2$$

(3) $\qquad\qquad ab - ba < r_2 s_2 - r_1 s_1.$

From (2) it follows:

(4) $\quad r_2 s_2 - r_1 s_1 = s_2(r_2 - r_1) + r_1(s_2 - s_1) < 2c\delta/2c = \delta.$

(3) and (4) give $ab - ba < \delta$, which contradicts the definition of $\delta$. Thus we have proved that $ab - ba > 0$ is impossible. As $a$ and $b$ are interchangeable, $ba - ab > 0$ is also impossible. Consequently $ab = ba$.

**Exercise.** Complete the proof for the case where $a$ and $b$ are not both positive.

Combining the two preceding theorems we find

**Theorem 7.6.3.** Every Archimedean ordered division ring is a field.

This theorem together with Th. 3.2.9 and Th. 3.4.4 yields

**Theorem 7.6.4.** $P_{10}$ is valid in $\mathfrak{P}(D_{11}, \; C1-C6, \; A)$.

The field $R^*(t)$ of formal power series over the field $R$ of rationals, as constructed above, is an example of a non-Archimedean ordered field, for in $R^*(t)$ we have $n \cdot 1 < t$ for every natural number $n$. Hence $\mathfrak{P}(R^*(t))$ is a model of $\mathfrak{P}(P_{10}, \text{C1}-\text{C6})$, in which A is not valid. This gives:

**Theorem 7.6.5.** A is independent of V1, V2, V3, $P_{10}$, C1—C6.

In the case where $D_{11}$ is valid, so that $\mathfrak{T}(\mathfrak{P})$ is a division ring $\mathfrak{F}$, we can give another geometric interpretation of the Archimedean axiom. Let us start with 4 points $O$, $X$, $Y$, $E$, of which no three are collinear. We consider the set of points which can be constructed from $O$, $X$, $Y$, $E$ by applying the operations of joining two points by a line and intersecting two lines a finite number of times. Note that it is not allowed to choose a point or a line at random; only previously constructed points may be connected by a line, and previously constructed lines may be intersected. The set of points obtained in this way is the *net* generated by $O$, $X$, $Y$, $E$. A line that connects two points of the net is called a line of the net.

**Theorem 7.6.6.** The net generated by $O$, $X$, $Y$, $E$, consists exactly of the points whose homogeneous coordinates in the coordinate system $OXYE$ belong to the prime field $\mathfrak{F}_0$, contained in $\mathfrak{F}$.

PROOF. The coordinates of $O$, $X$, $Y$, $E$ which are all 0 or 1, belong to $\mathfrak{F}_0$. If $P(p_0, p_1, p_2)$ and $Q(q_0, q_1, q_2)$ have coordinates in $\mathfrak{F}_0$, then the coordinates $\xi_0$, $\xi_1$, $\xi_2$ of $PQ$ must satisfy

$$p_0\xi_0 + p_1\xi_1 + p_2\xi_2 = 0 \quad \text{and} \quad q_0\xi_0 + q_1\xi_1 + q_2\xi_2 = 0.$$

As $\mathfrak{F}_0$ is commutative, the solution of these equations is $(p_1q_2 - p_2q_1, \ p_2q_0 - p_0q_2, \ p_0q_1 - p_1q_0)$, which again belongs to $\mathfrak{F}_0$. Similarly, if $l(\xi_0, \xi_1, \xi_2)$ and $m(\eta_0, \eta_1, \eta_2)$ have coordinates in $F_0$, then their point of intersection $(\xi_1\eta_2 - \xi_2\eta_1, \ \xi_2\eta_0 - \xi_0\eta_2, \ \xi_0\eta_1 - \xi_1\eta_0)$ has also coordinates in $\mathfrak{F}_0$. This suffices to prove that every point in the net has coordinates in $\mathfrak{F}_0$.

Conversely, in order to show that every point with coordinates $(p_0, p_1, p_2)$ in $\mathfrak{F}_0$ belongs to the net, we shall prove it first for every point $(1, p, p)$ on $OE$ with $p$ in $\mathfrak{F}_0$. As before, we denote the point $(1, p, p)$ by $p$. By the construction of $p + q$ and of $pq$ (§ 3.2, (9) and (10)), it is clear that if $p$ and $q$ belong to the net, then $p + q$ as well as $pq$ belong to the net. Given $p \neq 0$, we construct $p^{-1}$ as follows.

$$p^{-1} = \{(pX \cap EY)O \cap EX\} Y \cap OW.$$

Hence, if $p$ is in the net, so is $p^{-1}$.

Now any element of the prime field $\mathfrak{F}_0$ can be obtained from 1 by applying the operations of addition and inversion a finite number of times. Thus every point $p$ on $OE$, where $p \in \mathfrak{F}_0$, belongs to the net. Furthermore, if $P = (1, p_1, p_2)$ and $p_1$, $p_2$ belong the net, then so does $P = Xp_2 \cap Yp_1$. Finally, if $P = (0, p_1, p_2)$ and $p_1$, $p_2$ belong to the net, then we construct first $Q(1, p_1, p_2)$; both $Q$ and $P = OQ \cap XY$ belong to the net.

This completes the proof of Th. 7.6.6.

COROLLARY 7.6.6. In $\mathfrak{P}(D_{11}, C1-C6)$, the net generated by $O$, $X$, $Y$, $E$, consists of the points with rational coordinates in the system $OXYE$.

In fact, in this case $\mathfrak{F}$ is an ordered division ring so $\mathfrak{F}_0$ is the field of rationals.

**Theorem 7.6.7.** In $\mathfrak{P}(D_{11}, C1-C6, A)$, the net generated by four points, of which no three are collinear, is dense on any of its lines.

PROOF. Take two points of the line as $O$, $E$ of a coordinate system and choose $X$, $Y$ in points of the net. Now Th. 7.6.7 follows immediately from Th. 7.6.6 and Th. 7.6.1.

## § 7.7. The axiom of continuity.

The system of axioms V1, V2, V3, $D_{11}$, C1-C6, A is still not categorial, for we obtain non-isomorphic models for it by constructing projective planes over non-isomorphic Archimedean ordered fields, e.g. the field $R$ of rationals and the field $C$ of real numbers. Categoricity can be enforced by requiring that the field which is attached to the geometry shall be isomorphic to the field of reals. One of the methods by which this can be achieved is by the axiom of Dedekind. For the formulation of this axiom we need some set-theoretic notions.

Let $S$ be an ordered set; the order relation in $S$ is denoted by $<$. Let $V$ and $W$ be non-empty subsets of $S$ such that $S$ is the ordered sum of $V$ and $W$ (§ 1.2); then $V$ and $W$ determine a *cut*

$(V|W)$ in $S$. Hence if $(V|W)$ is a cut in $S$, then (i) $V \cup W = S$; (ii) if $v \in V$, $w \in W$, then $v < w$; (iii) $V \neq \emptyset$, $W \neq \emptyset$; (iv) $V \cap W = \emptyset$. Conditions (i), (ii), (iii) are necessary and sufficient for a cut; (iv) follows from (ii).

With respect to a cut $(V|W)$ in $S$ four cases are possible, namely

a)   $V$ has a last element, $W$ has a first element.

b)   $V$ has a last element, $W$ has no first element.

c)   $V$ has no last element, $W$ has a first element.

d)   $V$ has no last element, $W$ has no first element.

In case a) the cut $(V|W)$ determines a *jump* in $S$, in case d) it determines a *gap* in $S$, while in cases b) and c) the cut is *continuous*.

If $S$ is an ordered group, no cut in $S$ determines a jump. In fact, suppose that the cut $(V|W)$ determines a jump; let $a$ be the last element of $V$ and $b$ the first element of $W$, so that $a < b$. Then $a < (a+b)/2 < b$; the element $(a+b)/2$ can belong neither to $V$, nor to $W$, which is impossible.

We now introduce Dedekind's axiom De, which can be formulated in $\mathfrak{P}(C1–C6)$.

**Axiom De.** In $\mathfrak{T}(\mathfrak{P})$ no cut determines a gap.

**Theorem 7.7.1.** Axiom A is valid in $\mathfrak{P}(D_{11}, \text{C1}—\text{C6}, \text{De})$.

PROOF. Let $a$ be an element of $\mathfrak{T}$ such that $a > 0$. Let $V$ be the subset of $\mathfrak{T}$ which is defined as follows:
$x \in V$ if and only if there exists a natural number $n$ such that $na > x$; $V = \{x|(\exists n)(na > x)\}$. Define $W$ by $W = \mathfrak{T} \backslash V$; $W = \{x|(n)(na \leq x)\}$. We wish to prove that $W$ is empty. Suppose the contrary. If $x \in V$, $y \in W$, we can find $n$ such that $na > x$; then $x < na \leq y$. It follows that $(V|W)$ is a cut in $\mathfrak{T}$ as soon as $W$ is not empty. Because $\mathfrak{T}$ is an ordered division ring, the cut cannot determine a jump; according to De it cannot determine a gap. Hence it is continuous. Moreover, $V$ can have no last element $h$, for then there would be a natural number $n$ such that $na > h$, $(n + 1) a > na > h$, $na \in V$, which contradicts the hypothesis. Consequently $W$ has a first element $g$. $g - a < g$, so $g - a \in V$. By the definition of $V$ there is an integer $m$ such that $ma > g-a$. This gives $(m + 1)a > g$, $g \in V$, which is a contradiction. It fol-

lows that $W$ is empty and $V = \mathfrak{T}$; this is exactly the content of A.

**Theorem 7.7.2.** $\mathfrak{T}(D_{11}, \text{C1}-\text{C6}, \text{De})$ is isomorphic to the field of real numbers.

PROOF. By Th. 7.7.1 and Th. 7.6.1 the field $R$ of rationals is dense in $\mathfrak{T}$. To any element $\alpha$ of $\mathfrak{T}$ we associate the set $R_1(\alpha)$ of rational numbers less than $\alpha$; $R_1(\alpha)$ has no last element. If $R_2(\alpha) = \mathfrak{T} \setminus R_1(\alpha)$, then $(R_1(\alpha)|R_2(\alpha))$ is a cut in $R$. If $\alpha \neq \beta$, then $R_1(\alpha) \neq R_1(\beta)$; this follows from the fact that $R$ is dense in $\mathfrak{T}$. Conversely, let $(R_1|R_2)$ be a cut in $R$, where $R_1$ has no last element. Define $V$ as the set of those elements of $\mathfrak{T}$ which are less than some element of $R_1$:
$V = \{x | x \in \mathfrak{T} \,\&\, (\exists r)(r \in R_1 \,\&\, x < r)\}$. It is clear that $R \cap V = R_1$. $V$ has no last element, for if $a \in V \,\&\, r \in R_1 \,\&\, a < r$, then $a < (a+r)/2 < r$, so $(a+r)/2 \in V$ and $(a+r)/2 > a$. Let $W$ be $\mathfrak{T} \setminus V$, then $(V|W)$ is a cut in $\mathfrak{T}$. $(V|W)$ cannot determine a gap, because De holds. As $V$ has no last element, $W$ must have a first element $\alpha$. By the definition of $V$, $R_1(\alpha) = R \cap V = R_1$. Hence the correspondence between $\mathfrak{T}$ and the cuts $(R_1|R_2)$ in $R$, where $R_1$ has no last element, is one-to-one.

It is well-known that there is also a one-to-one correspondence between the cuts $(R_1|R_2)$ in $R$, where $R_1$ has no last element, and the real numbers. Hence we have established a one-to-one correspondence $\varphi$ between $\mathfrak{T}$ and the field $C$ of reals, such that $\varphi(r) = r$ for every element $r$ of $R$. Moreover, the definition of $\varphi$ is such, that for every $\alpha$ in $\mathfrak{T}$, $r < \alpha$ is equivalent to $r < \varphi(\alpha)$. From this fact it is easy to infer that $\varphi$ is an isomorphism between $\mathfrak{T}$ and $C$.

By this categoricity proof we have reached a natural bound for this book. However, it must be repeated, many questions in the field have not been touched. I may mention the theory of finite projective planes and spaces, the connections of projective geometry with topology, and the algebraic investigation of various classes of ternary fields. In the following books several of these subjects are treated in detail. (1) and (3) also contain a bibliography. (1) G. Pickert, Projektive Ebenen, Springer Verlag, 2nd ed., 1975. (2) Daniel R. Hughes and Fres C. Piper, Projective planes, Springer Verlag, 1973. (3) P. Dembrowski, Finite geometries, Springer Verlag, 1968.

# APPENDICES

## Appendix 1

Several proofs of Hessenberg's theorem were given since 1905, most of them incomplete because no account was taken of special cases. (See for example A. Seidenberg, Amer. Math. Monthly **83** (1971), pp. 190–192). The idea of the proof given in this book is very simple (Pappos $\rightarrow$ permutation proposition $\rightarrow$ weak permutation proposition $\rightarrow$ Desargues), but here also complications arise in special cases.

## Appendix 2

J. C. E. Dekker (J. Symbolic Logic **14** (1976), p. 399) made the remark that the uniqueness of the solution in (ix) follows from (viii).

PROOF. Suppose $m \neq m'$, $a_1 \neq a_2$ and

$$\Phi(a_1, m, n) = \Phi(a_1, m', n') = r_1,$$
$$\Phi(a_2, m, n) = \Phi(a_2, m', n') = r_2,$$

Then the equations

$$\Phi(a_1, y, z) = r_1, \qquad \Phi(a_2 y, z) = r_2$$

would lhave the solutions $(m, n)$ and $(m\ddot{\imath}, n\ddot{\imath})$ for $(y, z)$, which contradicts (viii).

## Appendix 3

Some information on terminology will be useful for further reading.

A *loop* is a set $S$ in which a binary operation (here denoted by $+$) is defined with the properties

(1) For every $a$ and $b$ in $S$ the equation $a + x = b$ has a unique solution for $x$;

(2) for every $a$ and $b$ in $S$ the equation $y + a = b$ has a unique solution for $y$;

(3) there is a unit element $e$ such that $e + a = a + e = a$ for every $a$ in $S$.

A ternary field $\mathfrak{T}$ is a loop with respect to addition and $\mathfrak{T} \setminus \{0\}$ is a loop with respect to multiplication.

A ternary field satisfying $\Phi(x, m, n) = xm + n$ is called *linear*. A linear ternary field with associative addition is a *Cartesian group*. A *quasifield* is a Cartesian group with both distributive properties:

$$x(y + z) = xy + xz; \qquad (x + y)z = xz + yz.$$

Hughes and Piper (Projective planes) call any set in which a ternary function is defined a ternary ring; they call a ternary field as defined in § 3.1 of this book, a *planar* ternary ring. Pickert uses for this notion the term *Ternärkörper*.

## Appendix 4

It is worth mentioning that a direct proof of the fundamental proposition, based upon $P_{10}$, was given by F. Schur in 1898 (See his „Grundlagen der Geometrie", 1909, sections 16, 17). I preferred the proof given in § 6,1 because of the interesting algebraic method.

## Appendix 5

Another method was given by H. J. Arnold, Abh. Math. Sem. Univ. Hamburg **45** (1976), pp. 3–60.

# INDEX

Numbers refer to sections